Reefs and Evaporites— Concepts and Depositional Models

STUDIES IN GEOLOGY NO. 5

REEFS AND EVAPORITES—
CONCEPTS AND DEPOSITIONAL MODELS

Edited by

James H. Fisher

Published by
The American Association of Petroleum Geologists
Tulsa, Oklahoma, U.S.A., 1977

Published November, 1977

Library of Congress Catalog Card No. 77-90427

ISBN: 0-89181-009-9

The AAPG staff responsible:

Ronald Hart, Project Coordinator
Sally B. Hunt, Production Coordinator
Nancy G. Wise, Production
Regina Landon, Production

Printed by Edwards Brothers, Inc.
Ann Arbor, Michigan, U.S.A.

Contents

Introduction

Most of the papers included in this volume were presented at the 1975 annual meeting of the Eastern Section, AAPG, held at East Lansing, Michigan. The Michigan basin has long been recognized as a classic area for the study of evaporite deposits. With the intensified exploration for petroleum associated with Silurian pinnacle reefs, Michigan has also been recognized as a classic region for the investigation of barrier and pinnacle reef complexes. There is an intimate interrelation between the reefs and the early Salina evaporites; the simultaneous study of these features could solve many of the problems involved in the controversy over deep versus shallow water evaporites. However, there is a danger in studying a few reefs or one or two evaporite formations and then drawing sweeping generalizations about all the Niagaran reefs or all Salina evaporites. Broad regional studies of both the Middle and Upper Silurian strata of the Michigan basin are imperative if we are to arrive at meaningful conclusions on the nature of reef growth and evaporite formation.

The inner portion of the Michigan basin is surrounded by a massive barrier reef some 500 to 600 ft (152.4 to 182.8 m) thick. Pinnacle reefs occur in a belt 10 to 15 mi (16 to 24 km) wide, extending from the steep inner face of the barrier reef toward the center of the basin. The central basin facies is a relatively thin carbonate averaging 60 to 80 ft (18.2 to 24.4 m) in thickness. The barrier reef is dolomite, the basinal facies is limestone, and the pinnacle reefs are mostly dolomite with a few of the more basinward ones being limestone. What created this regional pattern of dolomitization and when did it occur?

Other problems involve the age of the reefs themsleves. One group maintains that the reefs reached their maximum height and that all growth ceased by the end of the Niagaran. Others argue that reef growth continued during early Salina time and that the reefs developed concurrently with the lowermost Salina units, the A-1 evaporite and the overlying A-1 carbonate. Still another group believes in a cessation of reef development during A-1 evaporite deposition as a result of a presumed drop in sea level, followed by a rise in sea level and renewed reef growth during A-1 carbonate deposition. This has important implications regarding the magnitude of the fluctuations in sea level invoked by the adherents of shallow water evaporites. One possible solution to the problem of age determination is development of reefs of varying heights during the Niagaran. Then if sea level was lower during the time of A-1 carbonate deposition, additional reef growth could be substantial for the shorter reefs, less for the intermediate reefs, and totally lacking for the taller reefs. This might be an explanation for the different ages determined in localized studies.

The Michigan basin is frequently cited in evaporite studies as a classic example of a barred basin in which the barrier reef was the encircling medium. It is suggested that limited circulation from the open sea into the basin was maintained via several passes through the reef. One might raise the question of how effective a barrier a porous and permeable reef would be against the seepage of normal marine waters into the basin. However, assuming that passes are the major inlets by which normal sea water enters the basin during an evaporite cycle, what parameters do we encounter? The barrier reef to the east and west underlies Lakes Huron and Michigan. Since drilling has never been permitted in these areas, we have no information on reef passes in these regions. The drilling in the northern barrier reef has been inadequate to provide a clear picture, but two probable passes are present. The southern barrier reef has been densely drilled and can be outlined in some detail. Two major passes exist where the threshold or maximum reef thickness in the passes is 240 ft (73.2 m). The configuration of the reef is such that effective restriction is achieved only where the reef is 450 ft (106.6 m) thick or greater. Assuming that the barrier reef is Niagaran, those who advocate evaporative drawdown for the A-1 evaporite, though still maintaining a restricted basin with influx of normal seawater, have about 200 ft (60.9 m) of leeway on sea level fluctuation calculations. Many workers have stressed basin restriction by the encircling barrier reef, however, by the end of deposition of the A-2 carbonate, the barrier reef was completely buried and no longer affected sedimentation. Yet the thick salt beds of the B and F units were deposited in the subsiding Michigan basin. This questions the need for a barrier in relation to the Salina evaporites of Michigan.

A second large barrier reef, the Fort Wayne bank, has been identified in Indiana, paralleling the Kankakee platform. Droste and Shaver (this volume) maintained that reef growth in the Fort Wayne bank began in Niagaran time and was continuous in one place or another throughout most, if not all, of the Salina. Assuming even a modest paleoslope from the Kankakee platform to the depocenter of the Michigan basin, we are confronted with a sea hundreds of feet deep along the basinward face of the Michigan barrier reef. It would seem that the Michigan pinnacle reefs would be hard pressed to stay within even reasonable proximity of sea level. The problem is compounded when we examine the pattern of deposition of the evaporite in the southern part of the basin. The A-1 evaporite consists of 450 ft (106.6 m) of rather pure salt in the center of the basin. At the basinward edge of the pinnacle belt, the salt thins abruptly and within a few miles passes into a thin anhydrite. The basinward portion of the anhydrite is evenly-bedded, but approaching the foot of the barrier reef it becomes nodular, and at the foot of the reef, the limit of anhydrite deposition, the anhydrite commonly is brecciated. How do we explain this pattern of anhydrite deposition in waters on the order of 500 ft (152 m) in depth?

On the other hand, those who advocate evaporative drawdown and the development of even a modest sabkha environment in the area of the pinnacle reef belt, have their problems. Whether or not one accepts a Salina age for the Fort Wayne bank, we are still faced with major fluctuations in sea level between the evaporite and carbonate phases of A-1 and A-2 deposition.

Well information on the Silurian of Michigan is abundant but so are the geologic solutions based on these data. Regional studies of the entire Middle and Upper Silurian section (supplemented by detailed investigations of critical areas) are a must if we are to resolve the reef and evaporite controversy.

JAMES H. FISHER
Michigan State University
East Lansing, Michigan

August 12, 1977

Depositional Environments of Pinnacle Reefs, Niagara and Salina Groups, Northern Shelf, Michigan Basin[1]

JOHN M. HUH,[2] LOUIS I. BRIGGS, and DAN GILL[3]

Abstract In the northern shelf zone of the Michigan basin, pinnacle reefs of the Niagara and Salina Groups underlie the A-2 evaporite of the Salina Group. They consist of four developmental stages—biohermal, organic-reef, supratidal-island, and tidal-flat—in order of evolution. Rocks of the first two stages are within the Guelph Formation, Niagara Group, whereas rocks of the last two stages make up the lower units of the Salina Group.

Strata of the biohermal stage, composed predominantly of crinoidal and bryozoan allochems, initially developed in quiet water below wave base. The bioherm grew to wave base by the end of the biohermal stage. Wave-resistant tabulate corals and stromatoporoids of the organic-reef stage developed in the high-energy zone above wave base, in a normal marine environment.

Supratidal and intertidal environments of the supratidal-island stage are characterized by dominance of algal stromatolites. At the bottom and top of the section deposited during the supratidal-island stage, vadose caliche, extensive solution leaching, iron oxide, and flat-pebble conglomerate are indisputable evidence of subaerial exposure and two episodes of erosion. Fragments of algal flat-pebble conglomerate are in reef talus below the Salina A-O carbonate and A-1 anhydrite, indicating that the supratidal-island stage is older than the lowermost evaporites of the Salina Group.

The tidal-flat stage is recorded by carbonate rocks that cover tops of the pinnacle reefs; these beds were deposited contemporaneously with the upper A-1 carbonate (Ruff formation) of the off-reef section. The algal pelletal-wackestone lithofacies overlaps fine mudstone of the off-reef A-1 carbonate at the reef flanks, and unconformably overlies the finely laminated algal stromatolite of the supratidal-island stage.

Introduction

Relations among Niagaran reef carbonates and lower Salinan carbonates and evaporites in the Michigan basin are a stratigraphic problem. Upper Niagaran carbonates clearly underlie lower Salinan evaporites at some places. Niagaran carbonates largely were organic framework and detritus associated with reef-platform and reef-pinnacle growth, and the question remains whether reef growth had ceased throughout the Michigan basin when deposition of evaporites began.

In northern Indiana and Illinois, reefs continued to flourish during deposition of the lower Salina Group (Shaver et al, 1971; Shaver, 1974). On the other hand, the Belle River Mills pinnacle reef had reached maximal growth and was exposed to subaerial erosion and leaching before deposition of Salina A-1 sediments in the southeastern Michigan basin (Gill, 1973, 1977a, b). Conflicting opinions of numerous other geologists (Mesolella et al, 1974, 1975; Jodry, 1969; Sharma, 1966; Felber, 1964; Sloss, 1969) are based on deductive arguments from indirect geologic evidence or on nearly direct evidence from separate environmental complexes of the basin. Thus, outside the specific environmental niche represented by the Belle River Mills reef, and especially in the northern Michigan basin, the matter of stratigraphic relations among strata of the Niagara and Salina groups has been problematic. The purpose of this study was to determine the geology and stratigraphy of Niagaran reef-reservoir rocks in the northern pinnacle-reef belt, and to infer paleoenvironments of Niagara-Salina reef growth and evaporite deposition in the Niagara and Salina Groups.

[1] Manuscript received May 18, 1976; accepted November 16, 1976.
[2] Houston Oil and Minerals Corp., Houston, Texas 77002.
[3] Subsurface Laboratory, The University of Michigan, Ann Arbor, Michigan 48109.

This paper is a condensation and modification of the Senior author's doctoral thesis at The University of Michigan. We are indebted to C. I. Smith, R. V. Kesling, P. L. Cloke, and W. C. Bigelow, The University of Michigan, for review of the thesis, and to L. E. Ratliff, J. L. Fisher, and others for criticism of the manuscript. D. Gill's participation in this study was supported by N.S.F. grant EAR76-17410.

The area of study is the northern shelf of the Michigan basin, which extends from Mason County to Presque Isle County in the lower peninsula of Michigan (Fig. 1). Since 1969, more than 360 oil- and gas-producing Niagaran pinnacle reefs have been discovered on the northern shelf (Mantek, 1976). This study was concentrated on Kalkaska County (Kalkaska 21 field, Kalkaska Township), where several wells were drilled in off-reef facies, in addition to the wells in the reef.

Previous Work

The first definition of Upper Silurian subsurface stratigraphy in the Michigan basin was by Landes (1945). Other contributions to understanding regional stratigraphy and paleogeography of the Middle and Upper Silurian were by Cohee (1948), Evans (1950), Alling and Briggs (1961), Ehlers and Kesling (1962), Pounder (1962), Ells (1967), Burgess and Benson (1969), Huh (1973a, b), Briggs and Briggs (1974), Shaver (1974), Mesolella et al (1974, 1975), Meloy (1974), Gill (1975, 1977b), and Nurmi (1975). Stratigraphic relations of carbonate rocks of Niagaran reefs, and of strata of the lower Salina Group in the southeastern Michigan basin were studied by Felber (1964), Sharma (1961, 1966), Jodry (1969), Ells (1960, 1962, 1969), and Gill (1973, 1977a).

General similarities and marked dissimilarities in results of work by Gill, Jodry, Sharma, Mesolella et al, and Felber are reasons for the present study. The conflict of opinion involves time and mechanisms of reef growth and their association with periods when evaporites were deposited. Only Gill (1973) concluded that reef growth entirely preceded deposition of evaporites; Sharma and Mesolella et al believed that reef growth alternated with deposition of evaporites, whereas Jodry and Felber argued for complete contemporaneity. In this paper, a model to explain reef-evaporite relations in the northern Michigan basin is proposed.

General Stratigraphy of the Niagara and Salina Groups
Niagaran Group

Niagaran rocks (Middle Silurian) are exposed in the southern part of the upper peninsula of Michigan, in southwestern Ontario, northwestern Ohio, northern Indiana, northeastern Illinois, and eastern Wisconsin (Fig. 1). Niagaran strata are within 550 ft (165 m) of the ground surface in the southeastern and southwestern parts of Michigan. In the northern Michigan basin, depth to Niagaran rocks ranges from 3,800 ft (1,160 m) to more than 7,000 ft (2,130 m).

The Niagara Group comprises the Burnt Bluff, Manistique, Lockport, and Guelph[4] Formations, which are equivalent to the "Clinton," "White Niagaran," "Gray Niagaran," and "Brown Niagaran" in the informal terminology of the petroleum industry (Fig. 2). In the southern part of the Michigan basin, the Manistique Formation consists of about 20 ft of light-colored dolomitic carbonates and shales; in the northern Michigan basin, it is clean cherty dolomite that locally is thicker than 400 ft (120 m).

The whitish gray to brownish gray lower Lockport (White Niagaran) grades to gray in the upper Lockport, which is distinctive in fresh cores. The Lockport thins toward the center of the basin where it is hematitic red; toward the margins of the basin the Lockport is more than 300 ft (90 m) thick in the carbonate reefal bank (Fig. 3).

The Guelph Formation, uppermost unit of the Niagara Group, is equivalent partly to the Brown Niagaran of the off-reef facies (Fig. 2). The Guelph is made up of organic skeletal wackestones of the biohermal stage and boundstones of the organic-reef stage of the pinnacle reefs, it is brown micrite in the off-reef strata. Abundant stromatoporoids and coarse- to medium-grained dolomitic, arenitic wackestone are characteristic of the carbonate platform reefal bank (Meloy, 1974).

Pinnacle reefs are along the shelf area of the basin. Across the northern shelf, heights of reefs increase basinward from about 300 ft (90 m) along the front of the bank to more than 600 ft (180 m) along the basinal edge of the shelf slope (Mantek, 1973, 1976).

---

[4] The Guelph is considered to be a reef facies of the Lockport Formation, and thus throughout the Great Lakes region has no precise stratigraphic level.

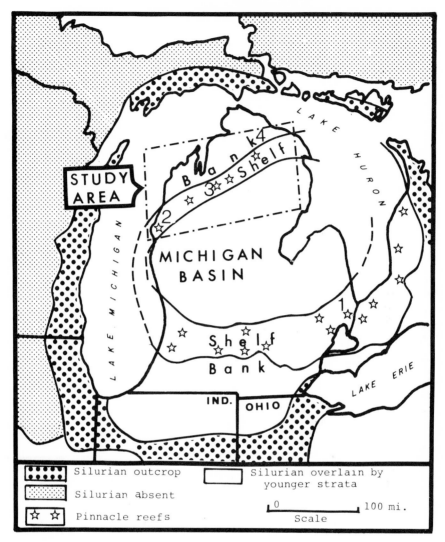

FIG. 1—Study area, distribution of Silurian rocks, reef banks, and pinnacle reefs: (1) Belle River
Mills reef, (2) Mason County, (3) Kalkaska County, and (4) Presque Isle County.

Recent exploration in northern and southern Michigan basin has provided more
accurate information about the Niagaran reef bank on the northern shelf and better
established development of the bank all around the basin (Fig. 1). The sedimentary
framework of Cayugan rocks involved development of Niagaran reef platforms and
formation of a physiographic barrier around the circumference of the basin. This
barrier prevented free interchange of brine between the open sea and the basin, and
controlled cyclic deposition of Cayugan evaporite and carbonate rocks (Alling and
Briggs, 1961).

Salina Group

The Salina Group (Upper Silurian) overlies the Niagara Group. At the tops of
pinnacle reefs, supratidal-island algal stromatolite is the lowermost unit of the Salina
Group. In basinal and inter-reef areas, the group consists of cyclic evaporites, limestones, dolomites and marls, divided into A-O carbonate, A-1 evaporite, A-1 carbonate

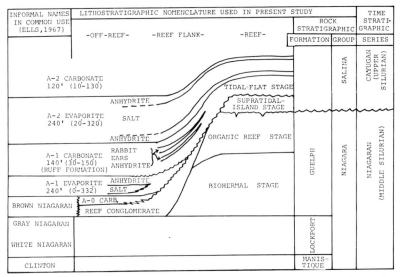

FIG. 2—Middle and lower Upper Silurian subsurface stratigraphic relationships and
nomenclature in the pinnacle reef belt, northern Michigan basin.

(Ruff formation), A-2 evaporite, A-2 carbonate, B-salt, C-shale, D-salt, E-unit (marl
and dolomite), F-salt, and G-unit (Landes, 1945; Evans, 1950; Ells, 1967; Budros and
Briggs, this volume).

Major Depositional Environments of the Niagaran Rocks

The Niagaran section is thinner than 150 ft (46 m) in the central part of the basin
(P.N. 20500, Sec. 28, T24N, R2E); it thickens gradually toward the margins of the
basin where it is thicker than 600 ft (180 m). Niagaran rocks record three major
lithotopes: carbonate platform with reefal bank, carbonate shelf with pinnacle reefs,
and basin (Fig. 1).

Carbonate platforms and reefal banks were built along the basin margins that subsided slowly. The organic framework of carbonate-platform bank reefs was restricted
largely to lateral growth, with skeletal allochems being deposited on a broad, gentle
slope. Outer parts of the carbonate banks mainly were skeletal coarse arenite and
stromatoporoids, which are characteristic of biohermal and biostromal deposits
(Meloy, 1974; Mesolella et al, 1974). Other framework organisms included *Favosites,
Halysites,* and *Coenites.*

Pinnacle reefs were built along the basin shelf, where subsidence was more pronounced. Due to the higher rate of subsidence, reefs built on the shelf appear to have
grown mainly vertically; they were 300 to 600 ft (90 to 180 m) high. Heights of
pinnacle reefs correlate with positions on the shelf area, and the higher reefs were on
the basinward part of the shelf. Pinnacle reefs in northern Michigan range from ¼ to ½
mi (0.4 to 0.8 km) wide and ½ to 1½ mi (0.8 to 2.4 km) long. Dipmeter data indicate
that reefs have flat tops and steep sides with slopes of 30° to 40° (Mantek, 1972,
personal communication). Several cores (P.N. 28773 and P.N. 28229) show dips of
30° to 40° in the upper flanks of pinnacle reefs.

The basinal environment included mostly debris of crinoidal carbonate mudstone
washed from the reefs and platform. Skeletal arenite of the shelf inter-reef zone gradually grades to fine argillaceous carbonate mudstone near the center of the basin; the
mudstone is characteristic of deeper-water deposition.

Geology of the Pinnacle Reefs

Pinnacle reefs in the northern Michigan basin show four major growth stages:
biohermal, organic-reef, supratidal-island, and tidal-flat. Major growth stages and inter-

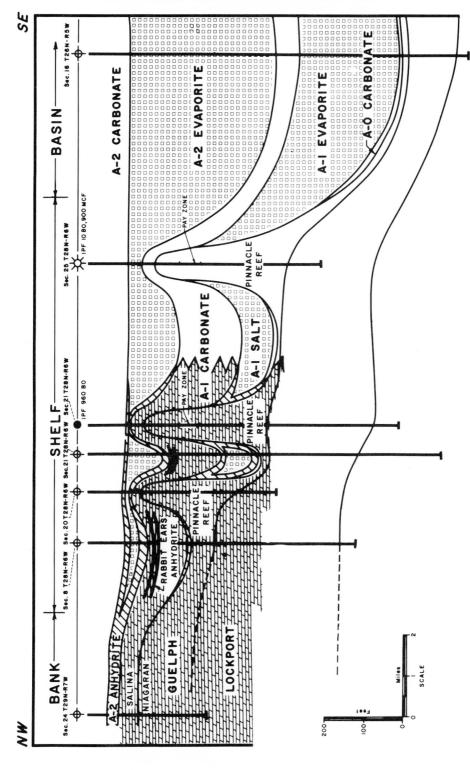

FIG. 3—Niagara-Salina stratigraphic relationships across the pinnacle reef belt, northern Michigan.

vening events are shown diagrammatically in Figure 4. Rock units of the first two stages compose the Guelph Formation of the Niagara Group. The succeeding supra-tidal-island stage (Figs. 2, 4D) marks initial deposition of the Salina Group, and the tidal-flat stage is represented by the A-1 carbonate of the Salina Group. Pinnacle reefs are flanked in the lower part by the reef-rubble conglomerate, which was deposited from debris of the organic reef of the Guelph Formation, and of the supratidal-island algal stromatolite (Fig. 5).

Biohermal Stage

Rocks of the biohermal stage form the lowermost part of the Guelph Formation, overlie the upper Lockport Formation (Gray Niagaran), and underlie rocks of the organic-reef stage (Figs. 2, 5). In ascending order, biohermal rocks are the skeletal biomicrite, biohermal-core, and skeletal-lithoclast facies. Skeletal biomicrite (Fig. 6) consists mainly of crinoids and bryozoa, which are mud-dwelling organisms (Kissling, 1969; Duncan, 1957). Thus, during deposition of skeletal biomicrite, water must have been so deep that wave action did not reach the substrate and remove clay-sized sediments. Deposition of the bioherms is interpreted to have begun in somewhat deep, quiet water, probably below wave base (Fig. 4A). Skeletal allochems and fine sediment were trapped by "forests" of living crinoids and bryozoa. Frame-building corals and tabular stromatoporoids invaded the mounds and built the biohermal-core facies on skeletal biomicrite. Tabular stromatoporoids have been interpreted as indicators of relatively deep, low-energy shoal environments in Devonian reef complexes of Alberta (Klovan, 1964; Murray, 1966; Leavitt, 1968; Fischbush, 1968).

The skeletal-lithoclast facies is developed along the top parts and flanks of bio-hermal mounds. Strata are composed of coarse skeletal fragments and lithoclast allochems. The coarser rocks indicate somewhat higher wave energy during deposition. Probably, the delicate crinoids and bryozoa could not survive in a high-energy environment, and a new faunal assemblage took their place. The skeletal-lithoclast facies records transition from the biohermal stage to the organic-reef stage, when tops of the bioherms reached wave base (Fig. 4B).

Organic-reef Stage

Rocks of the organic-reef stage overlie and flank strata of the biohermal stage; they underlie rocks of the supratidal-island stage of the Salina Group (Figs. 4c, 5).

Organic-reef rocks include the reef-core, reef-dwellers, and reef-detritus lithofacies, which are repeated in vertical section. Variations in lithofacies are due to periodic growth of frame-builders within reef cores and to deposition of intra-reef allochemical sediments. However, changes can be recognized in dominance of organisms during the organic-reef stage. *Labechiella* stromatoporoids were the more common early during deposition (Fig. 7), tabulate corals and the reef-dwellers assemblage were the more abundant in the middle part of deposition, and algal boundstone associated with massive stromatoporoids was predominant during the latter part of deposition.

The reef-core lithofacies consists mostly of frame-building stromatoporoids and tabulate corals (Fig. 7). Stromatoporoids were the major wave-resistant organisms and the essential frame-builders of the pinnacle reefs. According to Cloud (1952), massive stromatoporoids grew on solid substrates and were resistant to waves; they are evidence of high-energy environments. The reef-detritus facies is made up of debris broken and washed from the reef-core organisms.

The reef-dwellers lithofacies includes brachiopods, gastropods, rugose corals, crinoids, and bryozoa. These organisms existed throughout deposition of the organic-reef sediments. Such a community has been interpreted to have occupied a fore-reef ecologic position in the Upper Devonian Swan Hills reefs of Alberta (Jenik and Lerbekmo, 1968). Pinnacle reefs apparently were too small to permit development of lagoons in the crestal areas. In response to rapid subsidence, their vertical accretion and isolation prevented formation of a back-reef facies.

Great abundance of massive stromatoporoids and algal mudstone in the organic-reef lithofacies indicates that crests of reefs remained at sea level during construction of the

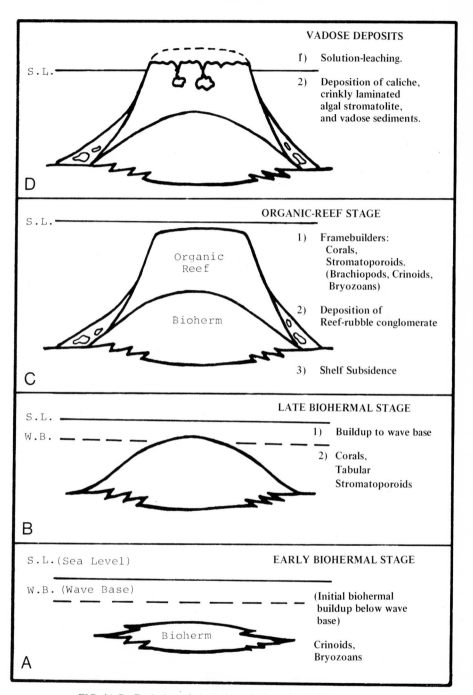

FIG. 4A-D—Evolution of pinnacle reefs, Salina and Niagara Groups.

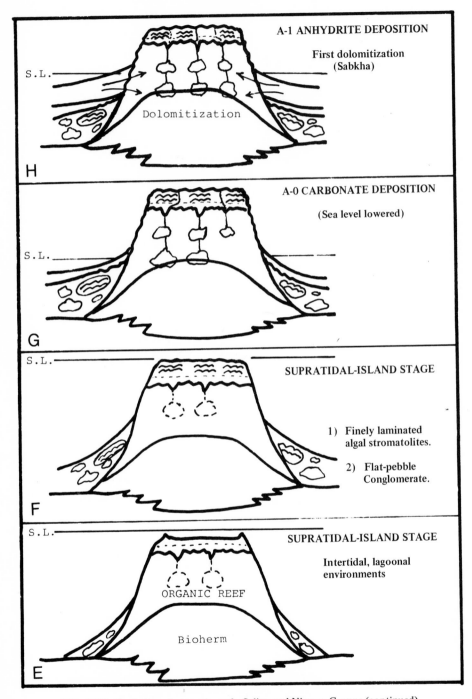

FIG. 4E-H—Evolution of pinnacle reefs, Salina and Niagara Groups (continued).

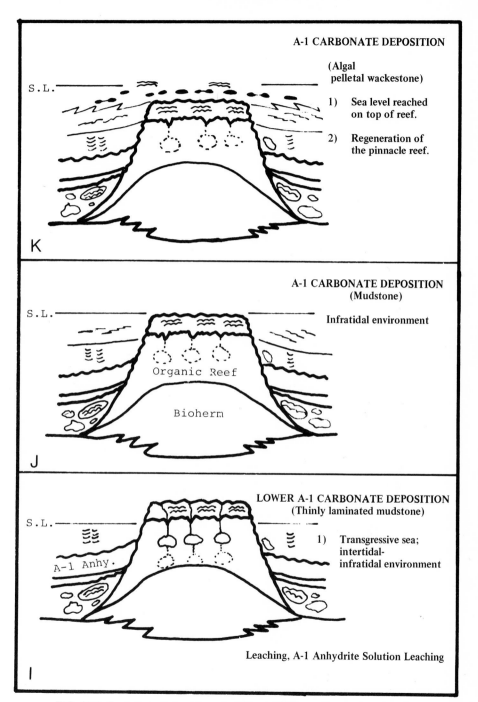

FIG. 4I-K–Evolution of pinnacle reefs, Salina and Niagara Groups (continued).

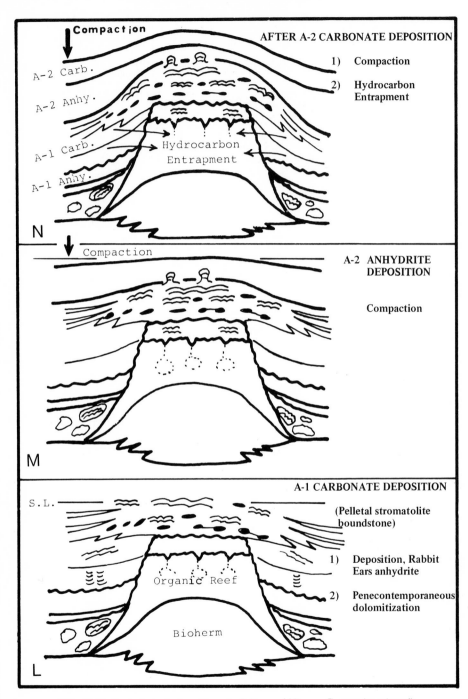

FIG. 4L-N—Evolution of pinnacle reefs, Salina and Niagara Groups (continued).

reefs. Rate of growth must have kept pace with subsidence of the basin and eustatic rise of sea level, so that at the end of the stage, tops of pinnacle reefs stood 300 to 600 ft (90 to 180 m) above the sea bottom.

Vadose Deposits

Vadose deposits are fibrous calcite linings, pisoliths, caliche and internal sediment in solution-leached vugs and cavities that are throughout rocks of the organic-reef stage (Figs. 4D, 8); however, they are especially abundant within 20 ft (6 m) of the top of the organic-reef section. Vadose rocks differ from organic-reef rocks, which commonly contain skeletal allochems, chiefly by having abundant insoluble residue. Fibrous calcite lined the solution-leached cavity walls (Fig. 8B), and vadose silt was deposited after crystallization of the fibrous calcite. Vadose pisoliths, laminated and pisolitic caliche, filled cracks, carbonate veins and other typical indicators of vadose karstic processes are also in the Belle River Mills reef in southeastern Michigan (Gill, 1972, 1973), and in pinnacle reefs of the northern Michigan basin.

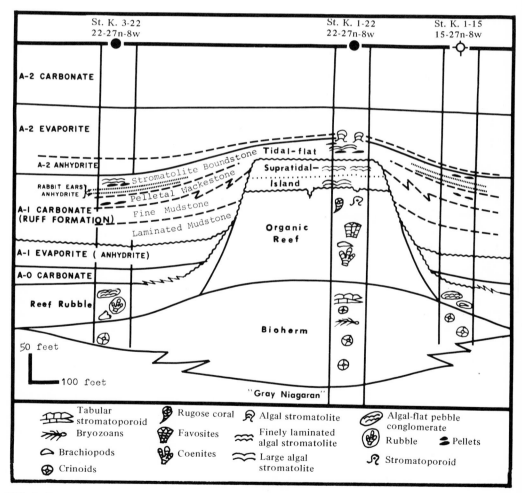

FIG. 5—Correlation of Niagaran pinnacle-reef units and Salinan units. Dashed line in supratidal-island section represents diastemic contact.

Supratidal-island Stage

Rocks of the supratidal-island stage unconformably overlie rocks of the organic-reef stage and are overlain disconformably by algal pelletal wackestone of the A-1 carbonate, Salina Group. In addition to boundary unconformities, which are recognizable both in the northern and southeastern Michigan basin, one or more diastemic contacts are within the lower part of the supratidal-island section.

Five lithofacies have been recognized in the supratidal-island rocks: algal stromatolite, algal-detritus wackestone, lagoonal mudstone, finely laminated algal stromatolite (LLH-c type), and flat-pebble conglomerate. Unlike the underlying biohermal and organic-reef rocks, these facies show persistent stratification.

The lowermost algal-stromatolite lithofacies comprises a crinkly, laminated algal stromatolite below (Fig. 8A) and a large, laterally-linked, hemispheroid algal stromatolite above. The crinkly stromatolite shows alternation of laminations and thin caliche deposits; of course, the caliche is evidence that crests of the pinnacle reefs were intermittantly exposed to subaerial conditions. The laterally-linked, hemispheroid, LLH-type algal stromatolite is an indicator of slight submergence and intertidal environments (Logan et al, 1964).

Algal-detritus wackestone unconformably overlies the hemispheroid algal stromatolite. It consists of coarse skeletal wackestone in the lower section and fine skeletal mudstone above. Predominance of skeletal allochems in the lower part indicates that the reefs may have been submerged to the shallow subtidal zone from the intertidal zone. Roehl (1967) also concluded that dominance of bioclasts, pellets, and muds is

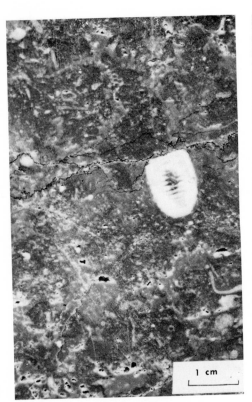

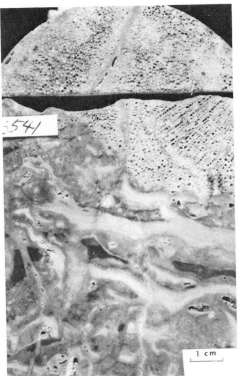

FIG. 6—Dark gray, skeletal dolomicrite of the biohermal stage, consisting mainly of crinoid fragments.

FIG. 7—Boundstone of the organic-reef stage, showing the stromatoporoid *Labechiella* and *Favosites* sp., which probably grew on the stromatoporoidal substrate.

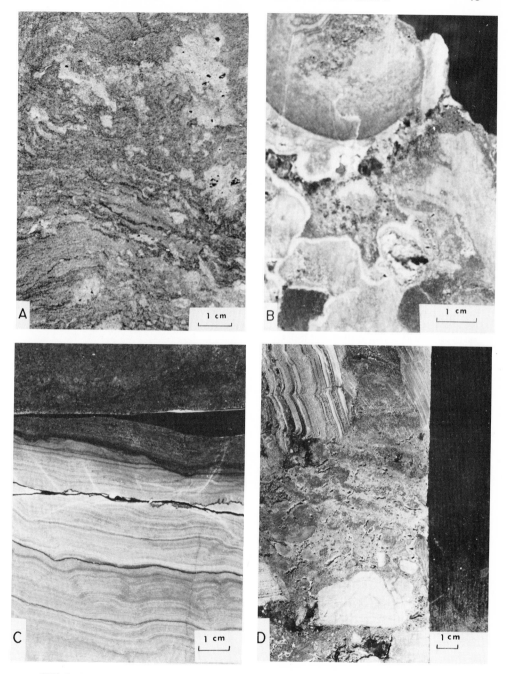

FIG. 8—Rock types of the supratidal-island stage.
 A. Crinkly laminated algal stromatolite, showing fine, wavy, crinkled algal laminae and light-colored caliche crust.
 B. Vadose lithofacies of the uppermost organic-reef stage, containing abundant solution-leached cavities with light-colored fibrous calcite lining (now dolomite), internal sediment (silt), and gilsonite (black).
 C. Finely laminated algal stromatolite (LLH-c type) at top of supratidal-island section, overlain by algal pelletal wackestone of the A-1 carbonate (Ruff formation) unit.
 D. Conglomerate of flat pebbles of finely laminated algal stromatolite.

evidence of the shallow subtidal zone. Fine skeletal mudstone of the upper section has extensive solution-leached texture, which indicates subaerial exposure.

The shallow lagoonal-mudstone lithofacies consists of fine massive mudstone in the lower section overlain by algal-laminated mudstone. The massive mudstone characteristically shows extensive bird's-eye texture. The rarity of fossils suggests a penesaline environment that would have inhibited organisms. Algal-laminated mudstone of the upper section contains mudcracks, anhydrite, and vadose sediments. These features indicate that the lagoons evolved to tidal flats (Fig. 4E).

The finely laminated algal stromatolite lithofacies (LLH-c type; Figs. 4F, 8C) overlies the lagoonal mudstone. The closely spaced, laterally-linked, hemispheroid algal stromatolite is identical to the LLH-type algal hemispheroids of Logan et al (1964) and Davies (1970), and is indicative of intertidal environments. Thus, tops of pinnacle reefs in the northern Michigan basin may have been in the intertidal zone during deposition of this lithofacies.

The flat-pebble conglomerate is composed of pebbles and cobbles of finely laminated algal stromatolite (Fig. 8D). These indicate early induration, possible desiccation, and storm-induced high energy in the high intertidal to supratidal environments. Furthermore, flat-pebble stromatolitic detritus in reef-flank conglomerate that underlies the A-1 anhydrite of the Salina Group demonstrates that algal stromatolite at the tops of pinnacle reefs predates the A-1 anhydrite. The supratidal sabkha environment deduced as being the environment of deposition of A-1 anhydrite, and the shallow-water to tidal-flat environment inferred for the A-1 carbonate lead inevitably to the conclusion that evaporative drawdown lowered sea level to below bases of pinnacle reefs after accumulation of the supratidal-island stromatolites, and before deposition of the oldest evaporites of the Salina Group.

Reef-rubble Conglomerate

The Reef-rubble conglomerate overlies the Lockport Formation and rocks of the biohermal stage; it underlies the A-O carbonate near flanks of pinnacle reefs, and the A-1 anhydrite (Fig. 2). The unit consists of a lower, coarse, lithoclastic conglomerate, a middle skeletal-lithoclast conglomerate, and an upper flat-pebble, algal-stromatolite conglomerate (Fig. 9A). The lower lithoclast conglomerate is composed of large, angular cobbles which contain brachiopods that inhabited the late biohermal and early organic-reef environments. The lower part of the conglomerate probably was deposited around pinnacle reefs from rubble produced by storm waves. The waves eroded and transported rocks from the lower part of the organic-reef section and the upper part of the biohermal section to bases of reefs.

Skeletal-lithoclast conglomerate is made up of clasts that contain skeletons of corals (*Coenites* and *Favosites*), crinoids, and brachiopods; the clasts are from deposits of the middle and late organic-reef stage. The conglomerate is interbedded with fine-grained arenite along flanks of reefs. Periodic storms and the intervening calm periods probably are the general conditions that induced the interbedding of conglomerates and arenites.

Pebbles of flat-pebble algal stromatolite are almost identical to flat pebbles that are within the LLH-c type algal stromatolites in the uppermost parts of the supratidal-island deposits (Fig. 9A, B). This similarity establishes provenance of the components, and relative ages of the flat-pebble algal stromatolite, the conglomerate, and evaporites of the Salina Group that overlie the conglomerate.

A-0 Carbonate

The A-O carbonate overlies, but partly grades into the reef-rubble conglomerate; both underlie the A-1 evaporite (Fig. 2). Thus, the A-O carbonate is the lowest unit of the Salina Group in the inter-reef and basinal regions. Brownish-gray color, thin, discontinuous, algal(?) and planar laminations interbedded with micrite, and the few ostracodes are characteristic of the unit. Algal(?) laminae and the predominantly fine-grained texture indicate deposition in relatively shallow water with little wave or current action. Sparsity and nondiversity of fossils suggest above-normal salinity. Transition upward to nodular and enterolithic (contorted) A-1 anhydrite supports the interpretation of high salinity and perhaps sabkha-type deposition near the end of deposition of A-O carbonate. If this environmental interpretation is correct, sea level must have lowered at least to the bases of pinnacle reefs after deposition of the A-O carbonate (Fig. 4G).

A-1 Evaporite

The A-1 evaporite is made up of anhydrite, halite, and sylvinite. The halite facies, commonly called the A-1 salt, is as thick as 475 ft (145 m) near the center of the basin (Ells, 1967); it thins gradually and pinches out at the bank-edge (Figs. 2, 3). The A-1 salt and anhydrite are considered to be contemporaneous by most subsurface geologists; the salt is judged to be predominantly the basin-center facies, and the anhydrite the facies marginal to the carbonate platform.

The A-1 anhydrite consists of three major rock types; in ascending order these are distorted nodular-mosaic anhydrite, distorted, laminated, massive-mosaic anhydrite, and nodular anhydrite (Fig. 10A, B). In spite of the three textural appearances of the anhydrite, all are nodular, and are mixed with brown dolomicrite.

Nodular anhydrite is deposited in the capillary zone of unconsolidated Holocene sediments in tidal flats and supratidal sabkhas, as along the Trucial Coast of the Persian Gulf. Nodular anhydrite is well-documented in the literature as an early penecontemporaneous diagenetic product of sabkha environments. For example, Murray (1964) stated that beds of nodular anhydrite result from growth of gypsum by displacement of soft sediment; thus the anhydrite must be considered as an early diagenetic facies, rather than a depositional facies. Murray also attributed ancient, bedded nodular anhydrite to deposition in supratidal flats. Features of the Upper Devonian Stettler Forma-

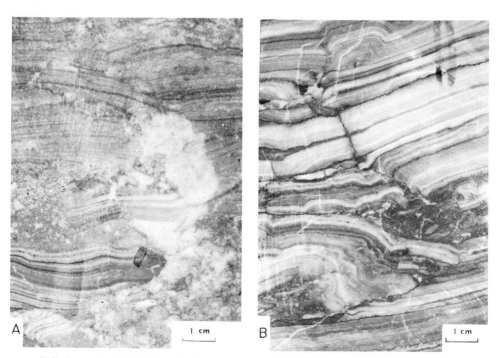

FIG. 9—A. Reef-rubble flat-pebble conglomerate composed of elongated cobbles and pebbles that show finely laminated algal-stromatolitic textures. Replacement anhydrite lath (dark rectangular crystal) is at lower central part of photograph.

B. Finely laminated algal stromatolite (LLH-c) of the supratidal-island stage. Note the exceptionally similar algal structures in Figs. 9A, B.

tion attributed to cyclic sabkha deposition (Fuller and Porter, 1969) are irregularly contorted and nodular anhydrite, with erosional surfaces at the tops of beds.

The A-1 anhydrite of the northern shelf of the Michigan basin is inferred to have been deposited under tidal-flat sabkha conditions, because it contains features similar to those of modern sabkhas and of the Stettler Formation. These features include nodular and enterolithic anhydrite, abundant algal mats and laminations, and evidence of leaching and erosion at the top of the anhydrite (Fig. 10B).

During deposition of the A-1 evaporite, the Michigan basin probably included (a) the outer zone of sabkha deposition (outer dolomitized pinnacle-reef belt), (b) the intermediate shelf zone of shallow-water salt deposition (inner limestone pinnacle-reef belt), and (c) the central, deeper-water zone of salt deposition (Fig. 4H).

A sharp unconformable contact at the top of the A-1 anhydrite is evidence of subaerial exposure. Some of the Reef-rubble conglomerate on the reef flanks also was reworked during this time. Exposure and erosion explain a thin or missing halite section above the sylvinite of the A-1 salt near the basin margin, and of a thick halite section above the sylvinite near the center of the basin (Matthews, 1970).

A-1 Carbonate (Ruff Formation) in the Reef Flank

The A-1 carbonate unconformably overlies the A-1 evaporite and underlies the A-2 evaporite in the reef-flank and inter-reef areas. Only the upper A-1 carbonate (algal pelletal wackestone) unconformably overlies crests of pinnacle reefs in the northern Michigan basin (Figs. 4K, 5).

The A-1 carbonate is as thick as 158 ft (48 m) along the outer margin of the shelf, near the edge of the carbonate platform. Thinning is abrupt across the pinnacle-reefs

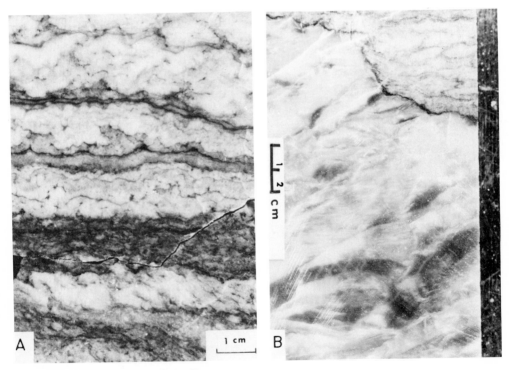

FIG. 10–A-1 anhydrite, Salina Group.
A. Distorted nodular-mosaic anhydrite interbedded with brownish-gray dolomicrite.
B. Distorted massive and mosaic anhydrite. Anhydritic nodular dolomicrite unconformably overlies distorted massive anhydrite. The unconformable contact may have resulted from leaching of anhydrite.

and gradual beyond, into the basin. The rock is principally dolomite on the carbonate platform and in strata of the outer marginal shelf; it is limestone in the pinnacle reefs and inter-reef sediments of the inner marginal shelf and basin. Four lithofacies have been identified adjacent to pinnacle reefs: thinly laminated mudstone overlain by fine mudstone, algal pelletal wackestone, and pelletal-stromatolite boundstone.

The thinly laminated mudstone contains a lower light brown dolomicrite and an upper dark gray, thinly laminated mudstone. The dolomicrite is made up of flat pebbles of finely laminated algal stromatolite in a brown micrite matrix. The large (2 cm x 5 cm) flat pebbles indicate occasional periods of high energy, such as storms, when the Reef-rubble conglomerate was reworked during transition from deposition of A-1 anhydrite to A-1 carbonate. The large flat-pebble lithoclasts and erosional contact at the top of the A-1 anhydrite (Fig. 10B) are supportive evidence for the inferred occasional periods of high energy.

The upper dark gray, thinly laminated mudstone contains partings of carbonaceous silt and shale, pyrite, and intraformational breccia and mudcracks (Fig. 11A). The mudcracks and breccia are evidence of subaerial exposure and occasional storms or changes of sea level.

Sparsity or absence of mudcracks and intraformational breccia higher in the section, and increase of carbonaceous laminae with argillaceous silts, clay, and pyrite, suggest that the depositional environment became stagnant, was protected from storms, and that clastic and organic matter were supplied in abundance. Likely environments for the above conditions would have been shallow lagoons or protected infratidal zones of pinnacle reef belts, analogous to Florida Bay behind the reef tracts.

Among lithofacies of the A-1 carbonate, this thinly laminated mudstone is the most likely source rodk for hydrocarbons, because of the relatively large amount of organic and carbonaceous matter.

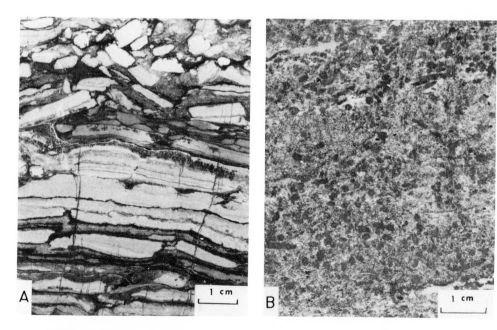

FIG. 11—A-1 carbonate (Ruff formation), Salina Group.

A. Thinly laminated mudstone showing intraformational breccia and desiccation cracks in the upper and lower parts, respectively. Pyrite (dark) is between thin laminated beds.

B. Algal pelletal boundstone composed of pellets and algal laminations. Pellets are 5 to 20 mm long, 1 to 2 mm wide, and are similar to modern fecal pellets.

Fine mudstone in the lower middle A-1 carbonate predominantly is brownish tan, thin-bedded, massive dolomicrite in the lower part and massive micrite with discontinuous horsetail-like laminae of carbonaceous seams in the upper part. The rare algal laminae probably are evidence of a depositional environment below the infratidal zone.

Bathurst (1971) reported a shallow subtidal, thin algal mat (2.5 mm) on muddy sand of Berry Islands in the Bahamas. However, grazing and burrowing organisms digest the mat and little trace of it is left in the subsurface (Bathurst, p. 123-126). A subtidal algal mat also was recorded at the Bimini Lagoon and the Little Bahama Bank (Scoffin, 1970; Neuman et al, 1970). Gradual increase and thickening of laminations in the upper fine mudstone are indicative of change from infratidal to low-intertidal environments. The fine mudstone is interpreted as having been deposited from below infratidal environments through low intertidal environments in a regressive sea.

Fecal pellets are distinctive of the overlying algal pelletal wackestone; they are well documented as being produced in the intertidal environment (Roehl, 1967; Cloud, 1962; Davies, 1970). Interfingering of algal pelletal wackestone and the underlying fine mudstone indicates alternations between intertidal and subtidal conditions during minor changes of sea-level.

The pelletal-stromatolite boundstone above (Fig. 11B) consists of planar algal stromatolite interbedded with enterolithic anhydrite, which commonly is called "Rabbit Ears anhydrite." Planar algal stromatolites are common in intertidal and supratidal environments of the Recent, and only rarely are developed in subtidal environments (Bathurst, 1971). Mats exposed to subaerial conditions are desiccated and form polygons. Absence of evidence of solution leaching and of desiccation cracks below the Rabbit Ears anhydrite indicates that the lowermost section of the pelletal stromatolite probably was deposited in the lower intertidal environment.

The Rabbit Ears anhydrite is characterized by nodular and enterolithic textures of the sabkha type. Both the desiccation cracks in boundstone between beds of anhydrite, and lithic properties of the Rabbit Ears anhydrite are evidence of subaerial exposure in a sabkha-type environment during deposition of the pelletal-stromatolite boundstone. Thus, during the latter part of deposition of A-1 carbonate, a broad tidal-flat to supratidal zone surrounded pinnacle reefs along the outermost shelf margin.

The predominantly dolomitized A-1 carbonate section below the upper Rabbit Ears anhydrite suggests that the sabkha environment of anhydrite deposition may be causally related to one stage of dolomitization of the pinnacle reefs and inter-reef sediments. Depositional environments of the overlying A-2 evaporite and the A-2 carbonate seem to have been similar to those of the A-1 units (Fig. 4L).

Summary

Several conflicting interpretations have been proposed to explain relationships between growth stages of pinnacle reefs and deposition of evaporites of the Salina Group. None of the models satisfies evidence found in this study. The writers propose the following:

1. During the biohermal stage, pinnacle reefs began to develop below wave base in quiet water. They grew into the zone of wave action. The predominantly micritic matrix, fossils comprising mostly mud-dwelling crinoids and bryozoans in the lower biohermal lithofacies, and skeletal lithoclasts in the uppermost lithofacies are evidence for this interpretation.

2. In the succeeding organic-reef stage, pinnacle reefs reached the high-energy zone above wave base, in a normal marine environment. Wave-resistant frame-building organisms, such as tabulate corals and stromatoporoids, indicate the high-energy environments. Tops of the reefs probably extended to sea level throughout the organic-reef stage. At the end of this stage, sea level dropped and upper parts of the reefs were exposed to subaerial erosion and leaching. Internal vadose sediments and caliche crusts with erosional surfaces in the uppermost part of the organic-reef section are evidence for this interpretation.

3. The supratidal-island stage began when the basin subsided or sea level rose to the reef crests; local intertidal and lagoonal environments resulted. The sea was restricted in circulation and penesaline, and a normal marine biota was absent. Deposition of supratidal-island rocks was terminated by major lowering of sea level. Subaerial exposure is recognized by much solution leaching, iron oxide, and by extensive erosion and formation of flat-pebble conglomerate in finely laminated algal stromatolites on tops of pinnacle reefs.

4. The A-O carbonate unit was deposited on the Reef-rubble conglomerate adjacent to the reefs, and as a facies of the conglomerate basinward in quiet water below wave base; algal(?) laminae and absence of normal marine fossils are characteristic of near-reef areas; nonalgal varved rocks are distinctive of inter-reef and basinal environments.

5. The A-1 anhydrite appears to have been deposited mostly in a sabkha environment. Thus, during deposition of A-1 anhydrite, the pinnacle reefs probably were completely exposed, as islands.

6. The thinly-laminated mudstone lithofacies of the lower A-1 carbonate was deposited in shallow intertidal to subtidal environments of a transgressive sea. Intertidal environments are indicated by the solution-leached upper surface of the A-1 anhydrite; reworked, finely laminated algal-stromatolite pebbles, and desiccation cracks in the lowermost part of the section are further evidence of intertidal environments. Absence of desiccation cracks, and abundance of dark, argillaceous carbonate mudstone in the upper part of the lower A-1 carbonate indicate shallow lagoonal or protected infratidal zones of the marginal shelf environment.

7. Fine mudstone of the lower middle A-1 carbonate was deposited in the infratidal environments of a continuously transgressive sea.

8. Algal pelletal wackestone of the upper middle A-1 carbonate unconformably overlies finely laminated algal stromatolite of the supratidal-island stage on pinnacle reefs (Fig. 5); the wackestone conformably overlies the fine mudstone in off-reef areas. Fecal pellets, burrows, and abundant algal mats, all of which are indicators of low-intertidal environments, are characteristic of the algal pelletal wackestone.

9. Pelletal-stromatolite boundstone of the upper A-1 carbonate conformably overlies the algal pelletal wackestone on the pinnacle reefs and reef flanks (Fig. 5). On reef flanks the boundstone is interbedded with the sabkha-type Rabbit Ears anhydrite. Solution-leached pellets and desiccation cracks in the planar algal stromatolite and sabkha-type anhydrite indicate prevalence of a broad tidal-flat environment late during deposition of the A-1 carbonate.

Deposition of algal pelletal wackestone and pelletal-stromatolite boundstone in the reef-crest and reef-flank regions constitutes the tidal-island stage of Huh (1973a, b). For the most part, these rocks were not additions to the reef framework. They are evidence of conditions quite unlike those of the organic-reef and supratidal-island stages, when reefs were 300 to 600 ft (90 to 180 m) above the sea floor. However, subsurface geologists commonly define tops of pinnacle reefs where A-2 anhydrite overlies A-1 stromatolitic carbonates. This practice seems to justify recognition and substantiation of the separate tidal-flat stage.

These sequential evolutions and overall sedimentologic context provide a framework within which the diagenetic and reservoir properties of the reefs can be explained satisfactorily. These properties include systematic partitioning across the pinnacle-reef belt in generation of secondary porosity, dolomitization, salt plugging, and in parallel partitioning of hydrocarbons within the reefs (Briggs and Briggs, 1975).

References Cited

Alling, H. L., and L. I. Briggs, 1961, Stratigraphy of Upper Silurian Cayugan evaporites: AAPG Bull., v. 45, p. 515-547.

Bathurst, R. G. C., 1971, Carbonate sediments and their diagenesis: American Elsevier, New York, 620 p.

Briggs, L. I., and D. Z. Briggs, 1974, Niagara-Salina relationships in the Michigan basin; in R. V.

Kesling, ed., Silurian reef-evaporite relationships: Michigan Basin Geol. Soc. Ann. Field Conf., p. 1-23.

———— 1975, Petroleum potential as a function of tectonic intensity in reef-evaporite facies, Michigan basin (abs.): AAPG-SEPM Ann. Meeting Abstracts, Dallas, Tex., p. 7.

Budros, R., and L. I. Briggs, 1977, Depositional environment of Ruff formation (Upper Silurian) in southeastern Michigan: (included in this volume).

Burgess, R. J., and A. L. Benson, 1969, Exploration for Niagaran reefs in northern Michigan: Ontario Petroleum Inst., 8th Ann. Conf., Tech. Paper 1, 30 p., 1969, p. 80-82; Dec. 29, 1969, p. 180-188; Jan. 5, 1970, p. 122-127.

Cloud, P. E., Jr., 1952, Facies relationships of organic reefs: AAPG Bull., v. 36, p. 2125-2140.

———— 1962, Environment of calcium carbonate deposition west of Andros Island, Bahamas: U.S. Geol. Survey Prof. Paper 350, 138 p.

Cohee, G. V., 1948, Thickness and lithology of Upper Ordovician and Lower and Middle Silurian rocks in the Michigan basin: U.S. Geol. Survey Oil and Gas Inv., Prelim. Chart No. 33.

Davies, G. R., 1970, Carbonate bank sedimentation, eastern Shark Bay, Western Australia: AAPG Mem. 13, p. 85-168.

Duncan, H., 1957, Bryozoans—annotated bibliography; in H. S. Ladd, ed., Paleoecology: Geol. Soc. Amer. Mem. 67, p. 783-799.

Ehlers, G. M., and R. V. Kesling, 1962, Silurian rocks of Michigan and their correlation; in Silurian rocks of the southern Lake Michigan area: Michigan Geol. Survey, Michigan Basin Geol. Soc. Ann. Field Conf. 1962, p. 1-20.

Ells, G. D., 1960, Silurian oil and gas developments in Michigan: Interstate Oil Compact Comm. Bull., v. 2, p. 64-78.

———— 1962, Silurian rocks in the subsurface of southern Michigan; in J. H. Fisher, chmn., Silurian rocks of the southern Lake Michigan area: Michigan Geol. Survey, Michigan Basin Geol. Soc. Ann. Field Conf., p. 39-49.

———— 1967, Michigan's Silurian Oil and gas pools: Michigan Geol. Survey Rept. Inv. 2, 49 p.

———— 1969, Architecture of the Michigan basin: Mich. Basin Geol. Soc. Ann. Field Excurs., p. 60-88.

Evans, C. S., 1950, Underground hunting in the Silurian of southwestern Ontario: Geol. Assoc. Canada Proc., v. 3, p. 55-85.

Felber, B. E., 1964, Silurian reefs of southeastern Michigan: Unpub. PhD thesis, Northwestern University, 194 p.

Fischbush, N. R., 1968, Stratigraphy, Devonian Swan Hills reef complexes of central Alberta: Bull. Canadian Petroleum Geol., v. 16, p. 446-587.

Fuller, J. G. C. M., and J. W. Porter, 1969, Evaporite formations with petroleum reservoirs in Devonian and Mississippian of Alberta, Saskatchewan and North Dakota: AAPG Bull., v. 53, p. 909-926.

Gill, D., 1972, Karstic diagenesis in Niagaran Guelph reefs, Michigan, and the origin of stromatactis: Geol. Soc. America, Abstracts with Programs, v. 4, no. 7, p. 721-722.

———— 1973, Stratigraphy, facies, evolution and diagenesis of productive Niagaran Guelph reefs and Cayugan sabkha deposits, the Belle River Mills gas field, Michigan basin: Unpub. PhD thesis, Univ. Michigan, 275 p.

———— 1975, Cyclic deposition of Silurian carbonate and evaporites in the Michigan basin, discussion: AAPG Bull., v. 59, p. 535-538.

———— 1977a, The Belle River Mills gas field: productive Niagaran reef encased by sabkha deposits, Michigan basin: Michigan Basin Geol. Soc. Spec. Pub. No. 2, 188 p.

———— 1977b, Salina A-1 sabkha cycles and the Late Silurian paleogeography of the Michigan basin: Jour. Sed. Petrology (in press).

Huh, J. M., 1973a, Geology and diagenesis of the Niagaran pinnacle reefs in the northern shelf of the Michigan basin: Unpub. PhD thesis, Univ. Michigan, 253 p.

———— 1973b, Stratigraphy and diagenesis of Niagaran pinnacle reefs (Silurian) in northern Michigan basin (Abs.): AAPG Bull., v. 57, p. 785.

Jenik, A. J., and J. E. Lerbekmo, 1968, Facies and geometry of Swan Hills Reef Member of Beaverhill Lake Formation (upper Devonian), Goose River field, Alberta, Canada: AAPG Bull., v. 52, p. 21-56.

Jodry, R. L., 1969, Growth and dolomitization of Silurian reefs, St. Clair County, Michigan: AAPG Bull., v. 52, p. 957-981.

Kissling, D. L., 1969, Ecology of Silurian crinoid-root bioherms (abs.): Geol. Soc. America, Abstracts with Programs, pt. 7, p. 126.

Klovan, J. E., 1964, Facies analysis of the Redwater reef complex, Alberta, Canada: Bull. Canadian Petroleum Geol., v. 12, p. 1-100.

Landes, K. K., 1945, The Salina and Bass Islands rocks in the Michigan basin: U.S. Geol. Survey Oil and Gas Inv., Prelim. Map No. 40.

Leavitt, E. M., 1968, Petrology, paleontology, Carson Creek North reef complex, Alberta: Bull. Canadian Petroleum Geol., v. 16, p. 298-413.

Logan, R. W., R. Rezak, and R. N. Ginsburg, 1964, Classification and environmental significance of algal stromatolites: Jour. Geology, v. 72, p. 68-83.

Mantek, W., 1973, Niagaran pinnacle reefs in Michigan: Michigan Basin Geol. Soc. Ann. Field Conf., p. 35-46.

—— 1976, Recent exploration activity in Michigan: Ontario Petroleum Inst., Proc. 15th Ann. Conf., 29 p.

Matthews, D., 1970, The distribution of Silurian potash in the Michigan basin: Mich. Geol. Survey, 6th Forum, Industrial Minerals, p. 20- 33.

Meloy, D. U., 1974, Depositional history of the Silurian northern carbonate bank of the Michigan basin: Unpub. M.S. thesis, Univ. Michigan, 78 p.

Mesolella, K. J. et al, 1974, Cyclic deposition of Silurian carbonates and evaporites in Michigan basin: AAPG Bull., v. 56, p. 34-62.

—— 1975, Cyclic deposition of Silurian carbonates and evaporites in Michigan basin: Reply: AAPG Bull., v. 59, p. 538-542.

Murray, J. W., 1966, An oil producing reef-fringed carbonate bank in the Upper Devonian Swan Hills Member, Judy Creek, Alberta: Bull. Canadian Petroleum Geol., v. 14, p. 1-103.

Murray, R. C., 1964, Origin and diagenesis of gypsum and anhydrite: Jour. Sed. Petrology, v. 34, p. 512-523.

Neumann, A. C., C. D. Gebelein, and T. P. Scoffin, 1970, The composition, structure and erodibility of subtidal mats, Abaco, Bahamas: Jour. Sed. Petrology, v. 40, p. 274-297.

Nurmi, R. D., 1975, Stratigraphy and sedimentology of the lower Salina Group, Upper Silurian, in the Michigan basin: Unpub. PhD thesis, Rensselaer Polytechnic Inst., 261 p.

Pounder, J. A., 1962, Guelph-Lockport formations of southwestern Ontario: Ontario Petroleum Inst., 1st Ann. Conf. Sess. 5.

Roehl, P. O., 1967, Stony Mountain (Ordovician) and Interlake (Silurian) facies analogs of Recent low energy marine and subaerial carbonates, Bahamas: AAPG Bull., v. 51, p. 1079-2032.

Scoffin, T. P., 1970, The trapping and binding of subtidal carbonate sediments by marine vegetation in Bimini Lagoon, Bahamas: Jour. Sed. Petrology, v. 38, p. 249-273.

Sharma, G. D., 1961, Geology of the Peters Field, St. Clair County, Michigan: Unpub. PhD thesis, Univ. Michigan.

—— 1966, Geology of Peters Reef, St. Clair County, Michigan: AAPG Bull., v. 50, p. 327-350.

Shaver, R. H., 1974, Silurian reefs of northern Indiana; reef and interreef macrofaunas: AAPG Bull., v. 58, p. 934-956.

Shaver, R. H. et al, 1971, Silurian and Middle Devonian stratigraphy of the Michigan basin: a view from the southwest flank, in Geology of the Lake Erie islands and adjacent shores: Michigan Basin Geol. Soc. Ann. Field Excurs., p. 37-59.

Sloss, L. L., 1969, Evaporite deposition from layered solutions: AAPG Bull., v. 53, p. 776-789.

Sedimentology and Depositional Environments of Basin-Center Evaporites, Lower Salina Group (Upper Silurian), Michigan Basin[1]

ROY D. NURMI[2, 3] and GERALD M. FRIEDMAN[2]

Abstract Sedimentologic analysis of the lower Salina Group and distribution of the evaporite and carbonate lithofacies suggest that the Michigan basin was desiccated periodically. Deposition of the Salina Group began with lowering of sea-level at the end of the Niagaran Epoch. The marginal platform and pinnacle reefs were exposed subaerially during this first major Cayugan lowering of sea level, as shown by erosional features, weathering surfaces (siliceous crusts, clay seams), and diagenetic features of vadose origin in rocks directly below the Salina Group. In the basin-center area, Cayugan sedimentation began with subaqueous interstitial precipitation of lenticular gypsum crystals.

Sedimentologic evidence within the basic-center evaporites suggests that deposition of the first Cayugan evaporite unit began in the deepest water in the Michigan basin; the basin became progressively shallower by desiccation and infilling. The interpretation is supported by depositional analyses of Cayugan carbonate units.

Introduction

After more than 200 yrs of controversy, the subject of the origin of marine evaporites still evokes heated debate within the geological community. In the late 1700s the question of the origin of evaporites became entangled in the raging Neptunist-Vulcanist controversy. Werner and the Neptunist supporters considered the Zechstein saline deposits to be marine precipitates. The Vulcanists propounded a volcanic origin for these rocks, challenged Werner's work, and caused much of the geological community to reject the marine-precipitate origin of saline deposits. Of course, the controversy between a pluton-volcanic or a marine-precipitate origin has since been resolved. Only a few geologists, notably in the Soviet Union, still advance a Vulcanist position.

A uniformitarian approach to interpretation of evaporites began early in Germany. G. Bischoff "argued from his knowledge of the recent deposits in the Dead Sea and the North Caspian depressions that the salt-deposits within the earth's crust had taken origin in the same way from ancient basins of water as they became desiccated" (Von Zittel, 1901, p. 219). G. Bischoff also proposed that isolation of a lagoon could result in salts and detrital sediments infilling the lagoon. Although later geologists accepted the idea that evaporites could be deposited in a restricted lagoon, certain modifications of G. Bischoff's model were made to explain the origin of the ancient thick saline deposits (F. Bischof, 1875; Ochsenius, 1877). These geologists proposed a barred-lagoon model because they considered it unfeasible to flood and desiccate an isolated deep lagoon repeatedly, which would be necessary for accumulation of a thick body of evaporites.

A new controversy began when the barred-lagoon model was opposed strongly by Johannes Walther (1903), who invoked a desert origin to explain ancient, thick saline deposits. The Upper Silurian evaporites of Michigan became part of this controversy when Grabau and Sherzer (1910) interpreted evaporites of the upper Salina Group in southeastern Michigan as desert deposits. This interpretation was supported by Cook (1914). Nevertheless, Cole (1915), in a review of salt deposits in the Ontario part of the Michigan basin, stated that, "By far the greater number of writers on the geology of this district are in favour of the 'Bar' theory of formation."

Nearly all known evaporite models have been proposed for the origin of the basin-center Salina evaporites: a series of shallow-water basins (Branson, 1915), a partially enclosed epicontinental sea with central evaporation and centripetal currents (Krum-

[1] Manuscript received, July 12, 1976; accepted, March 21, 1977.
[2] Department of Geology, Rensselaer Polytechnic Institute, Troy, New York 12181.
[3] Present address: Schlumberger-Doll Research Center, P.O. Box 307, Ridgefield, Connecticut 06877.

SYSTEM	SERIES	ROCK-STRATIGRAPHIC UNITS	
UPPER SILURIAN	LOWER CAYUGAN	LOWER SALINA GR.	C SHALE
			B EVAPORITE
			A-2 CARBONATE
			A-2 EVAPORITE
			A-1 CARBONATE
			A-1 EVAPORITE
MIDDLE SILURIAN	NIAGARAN	GUELPH-LOCKPORT GR.	GUELPH FM
			LOCKPORT FM

STRATIGRAPHIC UNITS

FIG. 1–Stratigraphic terminology of the Middle Silurian and lower
Salina Group (Upper Silurian) of the Michigan basin.

bein, 1951; Sloss, 1953), a lateral salinity gradient across the basin (Scruton, 1953), subsidence of a saline epeiric shelf (Landes, 1960), a deep-water barred basin (Schmalz, 1969), a basinal density stratification (Sloss, 1969), a hypersaline-flat, or sabkha, environment (Friedman, 1972; Treesh and Friedman, 1974), and a desiccated basin (Nurmi, 1974).

Lithofacies

The lower Salina Group in the central Michigan basin includes three major salt units separated by carbonate units (Figs. 1, 2), which can be correlated reliably using wire-line-geophysical logs. The lowermost lithofacies described is the basinal laminated limestone-anhydrite lithofacies, which conformably overlies the Middle Silurian basinal carbonates. The uppermost lithofacies is the peloid-ooid lithofacies, which composes the basal part of the A-2 carbonate unit. The delineated lithofacies are included in the rocks of the A-1 evaporite, A-1 carbonate, A-2 evaporite, and basal A-2 carbonate units. Examination of the few available cores containing parts of the B evaporite and C shale (Fig. 1) indicates that these upper rock units were deposited in environments similar to those described for the B evaporite and C shale units in New York by Treesh and Friedman (1974). The fact that these units can be correlated by physical continuity from the Michigan basin across Lake Erie (Nurmi, 1974), eastward into central New York (Rickard, 1969) suggests an analogous depositional history throughout the Great Lakes region.

Lithofacies I: Laminated Limestone-Anhydrite

Description

Rocks that compose the laminated limestone-anhydrite lithofacies are basal strata of the Salina Group (Upper Silurian). This lithofacies conformably overlies limestones

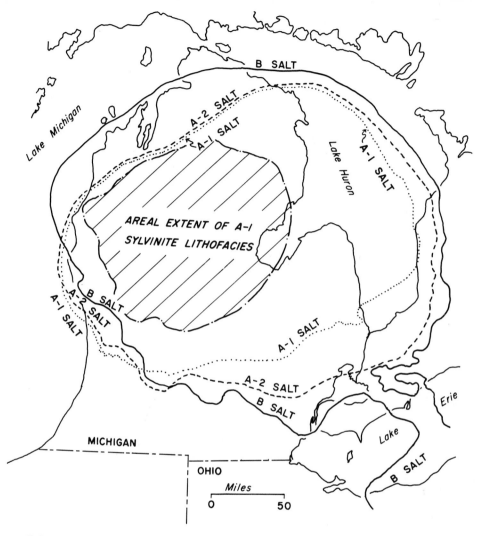

FIG. 2—Extent of salt units in the lower Salina Group. Note anomalous area in southwestern Michigan, where relative positions of the salt edges is reverse of the regional pattern. This is interpreted as the result of salt dissolution. Thickness and distribution of the A-1 sylvinite lithofacies is the result of greater subsidence in the northwest and inflow from the southeast.

of the Guelph-Lockport Group (Middle Silurian, Fig. 1). The lithofacies is restricted primarily to the central part of the Michigan basin, although it is present in the inter-pinnacle areas of the northwestern shelf (Fig. 3). The lithofacies is absent in the inter-pinnacle areas of the southeastern shelf. Areal distribution of the lithofacies corresponds to distribution of hematitic zones of the Middle Silurian basinal limestones.

In the center of the basin, there is less than 10 ft (3 m) of unfossiliferous laminated limestone between the laminated limestone-anhydrite lithofacies and the basinal limestones of the Guelph-Lockport Group (Niagaran). The upper laminites of this unfossiliferous limestone section contain lens-shaped aggregates of anhydrite crystals, which probably replaced lenticular gypsum. The lenses are parallel with laminae boundaries,

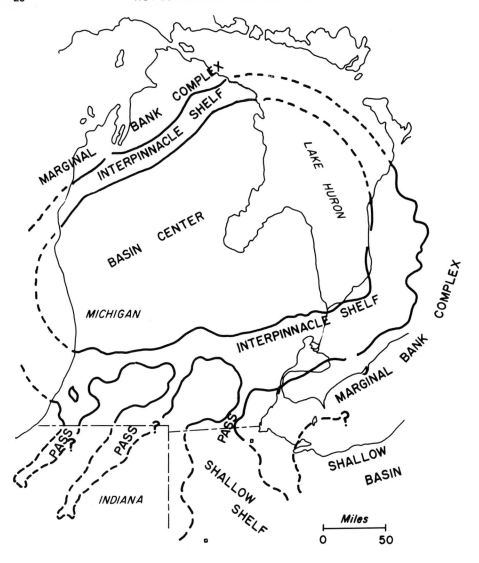

LATE NIAGARAN ENVIRONMENTS

FIG. 3–Paleogeography of the Michigan basin during late Niagaran time, immediately prior to the evaporative drawdown that resulted in deposition of the Cayugan A-1 evaporite unit.

and range in length from 70 to 675μ. Anhydrite crystals within the lenses range from 20 to 80μ long. Small laths of anhydrite (30μ long), randomly orientated, are present throughout the micrite laminae. In the basin-center core from Ogemaw County, Michigan (Fig. 4), lenses of anhydrite were observed in thin sections of a core sample of the rocks 5 ft (1.5 m) above a fossiliferous zone that contains abundant nautiloid cephalopods partly replaced by anhydrite. These lenses increase in number vertically in the micrite laminae until they are so abundant that anhydrite laminae are visible megascopically in the core (Fig. 5). Pyrite is common along boundaries of laminae.

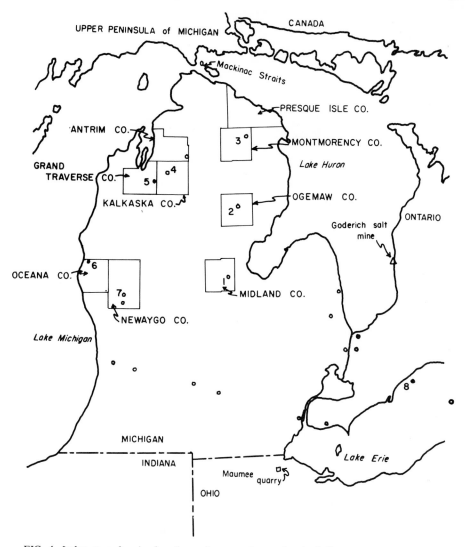

FIG. 4—Index map showing locations of cores and counties cited. Unnumbered locations represent cores studied but not cited. Listed are map location numbers with corresponding well names: **(1)** Dow 8 Chemical, **(2)** Amoco 1 Hunt Club, **(3)** Getty 1 Cain, **(4)** Pan American 1-19 State Kalkaska, **(5)** Pan American 1-14 State Union **(6)** Carter 12 Lauber, **(7)** Sun 4 Bradley, and **(8)** Consumers Gas 13009.

In the Dow 8 Chemical core from Midland County (Fig. 4) in the central part of the basin, the laminated limestone-anhydrite lithofacies is dark gray to black. The lithofacies on the margin of the basin-center area is much lighter in color; anhydrite laminae are gray, and micrite laminae are light to medium brown. Anhydrite laminae range in length from less than 0.1 cm to 2 cm.

Basinward of the pinnacle-reef trend, stringers of the laminated limestone-anhydrite lithofacies are interbedded with the basal part of the A-1 salt unit. Laminite stringers within the salt unit range from less than 1 in. (2.5 cm) to slightly more than 1 ft (0.3 m) thick. Salt sections between limestone-anhydrite stringers range from less than 1 in.

FIG. 5–Photomicrograph of a thin section, plane–polarized light.   Lenticular gypsum crystals replaced by anhydrite (white) are surrounded by micrite (dark). Coalesced lenses form the lower part of a megascopically visible anhydrite lamina (upper part of photomicrograph). Scale bar is 0.5 mm.

(2.5 cm) to more than 5 ft (1.5 m) thick. Petrographically, laminites within the salt unit are identical to laminites below the salt unit. However, in the Dow 8 Chemical core from the central part of the basin (Fig. 4), the uppermost lamina of the laminite stringers are contorted, and halite is under crests of some ptygmatic folds. In a core from the Getty 1 Cain, Montmorency County (Fig. 4), just basinward of the pinnacle-reef trend in northern Michigan, a series of laminite stringers nearly identical to that in the Dow Chemical core is present in the basal A-1 salt unit. In both cores the numbers of stringers and relative positions of each of the laminite stringers are the same. However the salt section is thicker in the Dow 8 Chemical core, in the basin center. The interval from the uppermost stringer to the base of the salt unit in the Getty 1 Cain core is 19.5 ft (5.9 m), whereas in the Dow 8 Chemical core the interval is 23.5 ft (7.16 m). In the Getty 1 Cain core, laminite stringers are thicker than in the Dow 8 Chemical core; also, there is a thick anhydrite cap at the top of each of the upper stringers, which is not found in the Dow 8 Chemical core. The anhydrite caps are partly leached and are filled by halite. In appearance, they are similar to fabrics above the laminite lithofacies on the northwestern shelf, where gypsum molds have been filled by halite. Thicknesses of anhydrite caps increase with each younger laminite stringer.

In the Pan American 1-19 State Kalkaska core from the interpinnacle area of the northwestern shelf (cf. Figs. 3 and 4), the top of the laminite sequence is capped by an anhydrite bed 5 in. (12.7 cm) thick. This bed contains well-preserved gypsum crystals that have been replaced by halite. These prismatic crystals generally are oriented vertically; some forms show swallow-tail twinning, and some crystals are longer than 1 in. (2.5 cm). The laminite sequence below this bed also contains irregularly shaped, halite-filled molds that may be gypsum crystals replaced by halite. Lower in the laminite sequence, zones of displacive nodular-mosaic to enterolithic anhydrite are interbedded with zones of limestone-anhydrite laminites.

In the southeastern area of the basin, just basinward of the interpinnacle shelf, enterolithic and nodular-mosaic anhydrite are interbedded with the laminated anhydrite-calcite lithofacies. Neither nodular anhydrite nor a modification of nodular anhydrite has been noted in cores of this lithofacies on basinal slopes or in the basin-center area.

Environment of Deposition

The laminated limestone-anhydrite lithofacies of the basin-center area differs significantly from that deposited in basin-marginal areas. Darkness of the basinal laminites and the odor of H_2S emitted during cutting of these laminites suggest that the laminites were deposited in an anaerobic environment. In cores from near the basin center, absence of evidence of gypsum crystals that might have formed at the sediment-water interface indicates that sulfate-reducing bacteria probably were active in the central basin. Pyrite along boundaries of laminae also suggests the former presence of sulfate-reducing bacteria. A basin-center anaerobic environment may imply that these basinal laminites were deposited at greater depths than laminites on the northwestern interpinnacle shelf or on the upper areas of basinal slopes.

The laminated limestone-anhydrite lithofacies presumably was deposited during concentration of the Niagaran sea that eventually resulted in deposition of halite and sylvinite. If the 50-m (164 ft) calculation for fictive depth of brine that precipitated the A-1 salt directly overlying this lithofacies is correct (Holser, 1966a), then brine in which the limestone-anhydrite lithofacies was deposited must have been somewhat deeper than 50 m (164 ft). Validity of the fictive-depth calculation, which is based on a bromide profile, has not been established conclusively. However, the regional relations between height of the Middle Silurian pinnacle reefs and thickness of the A-1 salt unit indicate a water depth of more than 150 ft (46 m) during deposition of the basin-center laminites (Fig. 6).

The small lenticular anhydrite forms within the calcite (micrite) laminae of this lithofacies are suggestive of interstitial gypsum crystals replaced by anhydrite. Similar lenticular gypsum crystals are growing interstitially in the intertidal sediments of these modern hypersaline lagoons: Abu Dhabi lagoon, Trucial Coast (Bramkamp and Powers, 1955); Laguna Madre, Texas (Masson, 1955); Shark Bay, Western Australia (Davies, 1970); and San Felipe, Baja California (Kinsman, 1969).

Although interstitial growth of lenticular gypsum does take place in intertidal environments, interstitial growth of these crystals does not dictate an intertidal origin. Probably the only requirement for growth of these crystals is increase in sulfate concentration of interstitial water. When the requisite concentration for precipitation of gypsum is reached, interstitial lenticular gypsum crystallizes. (In the intertidal-flat sediments of the Trucial Coast, the requisite concentration of approximately 65% is attained by evaporation of interstitial waters on the exposed mud flat [Butler, 1969].) Presumably, concentration of interstitial brine in the central part of the Michigan basin was the result of evaporation. Increase in temperature, salinity, or burial could have altered the original crystals to anhydrite.

It is suggested that at some point in concentration of sulfate ions in the water column, the rate at which sulfate ions diffuse into sediment will exceed the amount of organic matter needed by the sulfate-reducing bacteria to continue the cycle, or will

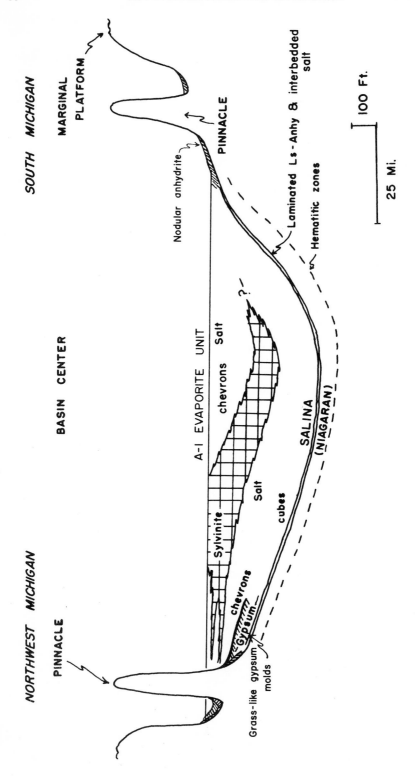

FIG. 6—North-south cross section of the A-1 evaporite unit, Michigan basin.

exceed the ability of bacteria to decompose the gypsum precipitated interstitially. Laboratory studies of Harrison and Thode (1958) and Kemp and Thode (1968) demonstrate that the rate of sulfate reduction is dependent upon the concentration of sulfate. This conclusion is supported by studies of Zobell and Rittenberg (1948), which showed that the population of sulfate-reducing bacteria in sediment decreases rapidly with increased depth. Sulfate-reducing bacteria at the sediment-water interface of anaerobic waters and in the uppermost sediment apparently have enough organic matter available to prevent accumulation of gypsum precipitated in the upper-water mass.

It is concluded that the laminated limestone of this lithofacies was deposited in the central part of the basin in water probably deeper than 50 m (or about 165 ft). During, and subsequent to deposition of the limestone, lenticular gypsum crystals were deposited interstitially. With burial they were dehydrated to anhydrite. Concentration of these crystals in layers, perhaps of high permeability, resulted in the megascopically visible anhydrite laminae of this lithofacies.

The nodular-mosaic to enterolithic anhydrites interbedded with the laminite lithofacies on the interpinnacle shelf in northwestern Michigan, and on the upper basin-center slopes in southeastern Michigan, may have developed during periods when these areas were exposed subaerially. Concentration of interstitial fluids by a process such as "capillary concentration" (Friedman and Sanders, 1967) could result in sulfate concentrations high enough to precipitate primary anhydrite, which causes the displacive nodular habit of the anhydrite (Kinsman, 1969; Bush, 1973). The cyclic development of laminites and nodular-mosaic enterolithic anhydrite could have been due to minor fluctuations of sea level that exposed periodically the subaqueously deposited laminites. However, the possibility that displacive anhydrite could have formed subaqueously should not be dismissed simply because no modern analog has been found. Only a decade ago, primary displacive anhydrite was discovered in sabkha deposits of the Trucial Coast (Shearman, 1963). In order to explain the displacive-nodular anhydrite forms in the sabkha in Abu Dhabi, Kinsman (1969) and Bush (1973) suggested that primary anhydrite precipitated upon anhydrite nuclei that resulted from the dehydration of the gypsum. A similar process, but one occurring subaqueously, can be envisioned to account for the displacive nodular-mosaic to enterolithic anhydrite zones in the laminated limestone-anhydrite lithofacies of the Salina Group.

Lithofacies II: "Grasslike" Gypsum Molds and Laminated Anhydrite

Description

The "grasslike" gypsum molds and laminated anhydrite of lithofacies II are present in the lower part of the A-1 salt unit on the northwestern interpinnacle shelf of Michigan. This lithofacies is composed of alternating layers of laminated anhydrite and vertically oriented gypsum-crystal molds (Fig. 7). The crystal molds are believed to be primary gypsum crystals either filled or replaced by halite (Fig. 7). The crystal molds are elongated vertically and are irregularly six-sided when viewed in horizontal cross section. Similar but larger molds in the same core have swallow-tail, twinned, upper terminations. Crystal molds in this lithofacies generally are arranged normal to the subjacent lamina, but developments with a slightly radial orientation also are seen. Lengths of these pseudomorphs are approximately 1.25 mm on the average. Similar pseudomorphs, as long as 30 mm (1.2 in.) are at the top of the laminated limestone-anhydrite lithofacies, which lies 18 ft (5.5 m) below the grasslike gypsum-molds and laminated-anhydrite lithofacies.

The interpretation that these crystals are pseudomorphous after gypsum is based on the crystal form, swallow-tail terminations, vertical growth position, interbedding of these crystals with finely crystalline laminated anhydrite that probably replaced gypsum, partial replacement of the grasslike crystals by anhydrite, and comparison with both modern and ancient gypsum deposits. Very similar vertically oriented or grasslike gypsum crystals, unreplaced by anhydrite, were described in modern lagoons, salt lakes, solar salt ponds, and in the upper Miocene Gessoso-Solfifera and Pleistocene Montallegro Formations of Sicily (Schreiber and Kinsman, 1975). The term "grasslike"

FIG. 7—Photomicrograph showing a thin section, in plane-polarized light. Two crystal molds contain halite and anhydrite, presumed to have replaced prismatic gypsum crystals. White areas are halite; finely crystalline material within halite is anhydrite. Anhydrite lamina that crosses the two crystal molds can be traced across the entire thin section. These crystals are interpreted as having grown at the sediment-water interface. The anhydrite lamina appears to have buried a layer of gypsum crystals; later, the crystals continued to grow in crystallographic continuity above the lamina. The thin section is from a "grasslike" gypsum-molds and laminated-anhydrite lithofacies. Scale bar is 0.5 mm.

was applied by Richter-Bernburg (1973) to the vertically oriented gypsum crystals in the upper Miocene Gessoso-Solfifera Formation. Similar crystal forms also replaced by anhydrite and halite were interpreted as pseudomorphous after gypsum in the Upper Permian Salado Formation in New Mexico (Schaller and Henderson, 1932), in the Permian Wellington Formation in Kansas (Jones, 1965), and in the Permian York-shire evaporites in England (Stewart, 1951).

Environmental Interpretation

The grasslike gypsum-molds and laminated-anhydrite lithofacies of the A-1 evaporite unit seems to be of shallow-water origin. Deposits resembling this lithofacies were reported in shallow, modern hypersaline lagoons along the Trucial Coast of the Persian Gulf (Shearman, 1971; Fairbridge, in Schreiber, 1974), in shallow, saline Marion Lake, South Australia (Hardie and Eugster, 1971), and in solar salt ponds in Newark, California (Schreiber and Kinsman, 1975). All primary gypsum crystals have been cited either from evaporite formations considered to be of shallow-water origin or from marginal areas of an evaporite basin. In the lower Salina Group this lithofacies has been found only in a marginal area of the A-1 evaporite unit. Neither grasslike nor large gypsum crystals were formed in deeper-water environments of the late Miocene of the Mediterranean basin (B. C. Schreiber, personal communication, 1975). Of interest is the fact that no large primary gypsum crystals have been reported in the Castile Formation in Texas and New Mexico, a formation reputed by many geologists to have been deposited in a deep-water, deep-basin environment (King, 1947; Anderson et al, 1972). Examination of modern and ancient deposits similar to the grasslike gypsum beds suggests that the gypsum is formed in shallow-water environments. A shallow-water origin for this lithofacies is supported further by its association with nonplanar stromatolite structures, flat-pebble conglomerates, and nodular anhydrite.

Grasslike gypsum crystals in the A-1 evaporite unit are thought to have crystallized at the sediment-water interface, whereas the lenticular gypsum crystals in the basal laminites crystallized interstitially. D. J. Shearman (personal communication, 1974) states that gypsum crystals growing at the sediment-water interface rarely are of lenticular habit. The vertically oriented grasslike gypsum crystals have been shown to develop at the sediment-water interface in modern lagoons and salt ponds (Schreiber and Kinsman, 1975). In thin sections of crystal molds of the grasslike gypsum-molds and laminated-anhydrite lithofacies, single laminae of finely crystalline anhydrite can be traced across the thin section (Fig. 7) and are seen crossing adjacent crystal molds. It appears that while the crystals were growing at the sediment-water interface, they occasionally were buried by a layer of gypsum, possibly precipitated from the water mass, and then they continued to grow above this layer in crystallographic continuity (Fig. 7). Corresponding observations from both modern and ancient gypsum deposits confirm an origin at the sediment-water interface for these former gypsum crystals. Whereas the physicochemical conditions of the basin-center environment were not conducive to crystallization at the sediment-water interface, it appears that the physicochemical conditions of the early Cayugan marginal environment were.

The conclusion is drawn that the grasslike gypsum molds and laminated anhydrite of lithofacies II were deposited in shallow, oxygenated waters. Absence of gypsum crystal forms in the central part of the basin, which would be suggestive of precipitation at the sediment-water interface, probably is due to degradation of sulfate minerals by anaerobic sulfate-reducing bacteria. An interpretation of a shallow-water origin of this lithofacies is strengthened by presence of domal stromatolites, flat-pebble conglomerates, and current structures in the underlying salt beds. Also of support is the interpretation of a shallow-salt-flat depositional environment for the overlying salt section, which is composed of chevron-zoned halite facies.

Lithofacies III: Nodular Anhydrite

Description

The nodular-anhydrite lithofacies is a marginal facies of each evaporite unit of the lower Salina Group. This lithofacies is present locally in the middle of the A-1 carbonate unit and is interbedded within the A-2 salt, A-2 carbonate, B evaporite, and C shale units. Within the A-1 evaporite unit, the nodular-anhydrite lithofacies is found only in the basinal depression—not on the subaerially exposed pinnacle reefs or marginal-bank complexes. It is best-developed on the southern interpinnacle-shelf area of the basin. Nearly all the forms of anhydrite described by Maiklem et al (1969) can be identified in the nodular-anhydrite lithofacies of the A-1 anhydrite unit; included are nodular, nodular-mosaic, mosaic, and massive anhydrite, and transitions among these forms. These anhydrite forms generally are both bedded and nonbedded varieties. A less-extensive nodular A-1 anhydrite lithofacies is on the northwestern side of the basinal depression. It is developed only on slopes of the subaerially exposed pinnacles and marginal-bank complex. On both sides of the basin, the nodular-anhydrite lithofacies is overlain unconformably by the A-1 carbonate.

The nodular-anhydrite lithofacies within the A-1 carbonate unit generally is referred to as the "Rabbit Ears" anhydrite. The anhydrite commonly is nodular, but nodular-mosaic and mosaic forms are present also. Areal extent of the Rabbit Ears anhydrite in southeastern Michigan was investigated by Budros (1974), who found that the nodular anhydrite is restricted to slopes of the pinnacle reefs. Huh (1973) observed that in northwestern Michigan, the nodular anhydrite similarly is restricted to margins of the pinnacles. A nodular anhydrite considered to be coeval with the Rabbit Ears anhydrite is present within the A-1 carbonate unit on the southern Ontario marginal platform.

The nodular-anhydrite lithofacies also is developed at the top of the A-1 carbonate unit throughout much of the basin. It has been deposited on the marginal platforms, pinnacle reefs, and interpinnacle-shelf areas of both the northwestern and southern parts of the basin. It is found on the upper slope of the basin-center area in the Goderich salt mine in Huron County, Ontario (Fig. 4). In Michigan, anhydrite developed in the uppermost beds of the A-1 carbonate unit is termed the "A-2 anhydrite unit." Forms of this anhydrite include nodular, nodular-mosaic, mosaic, and massive anhydrite. Like the A-1 anhydrite unit, the A-2 anhydrite commonly contains transitions among these forms, which are bedded and nonbedded. In the interpinnacle-shelf areas and on the upper basinal slopes, the nodular-anhydrite lithofacies is overlain by the A-2 salt unit. Exposure of the transition from A-1 carbonate to nodular anhydrite to A-2 salt in the Goderich salt mine shows an erosional surface at the top of the A-1 carbonate, where the nodular-anhydrite lithofacies is developed. Almost all cores of the nodular-anhydrite lithofacies contain abundant evidence of nonplanar and(or) planar stromatolites. Commonly, the growth form of the stromatolite is preserved where the stromatolite lies directly below a salt. Halite replacement and(or) filling of the stromatolite laminae are the reasons for good preservation. Beds of nodular anhydrite also are present within the A-2 salt unit in the interpinnacle areas.

On the Ontario marginal platform, nodular anhydrite is included in anhydrite beds within the A-2 carbonate. An example of such an anhydrite bed is shown in Figure 8. The nodular anhydrite is in the lower part of the anhydrite section. Higher in the core, a vertical succession of anhydrite forms range from distorted nodular anhydrite to distorted nodular-mosaic to bedded-mosaic anhydrite.

Environmental Interpretation

The origin and development of nodular anhydrite were the subject of many conflicting hypotheses until Shearman (1963) found this anhydrite forming in modern supratidal sediments of the Trucial Coast, Persian Gulf. Withington (1961) had described nodular calcium sulfates from ancient evaporite deposits and had deduced correctly that they grew in-place in the bottom sediments. He further maintained that this growth took place after the sediments were deposited but before they were lithified. However, Withington was unable to determine whether anhydrite or gypsum had formed the displacive nodules. Recent studies of sabkhas of the Persian Gulf substantiate that much of the massive and nodular anhydrite within the sabkha sediments resulted from the replacement of gypsum-crystal mushes (Kinsman, 1974). However, the contorted, layered, and coarsely nodular anhydrite within the prograding supratidal unit of nonmarine sediment was thought not to have a gypsum precursor; this anhydrite was considered to be primary.

On the marginal platform in the Michigan basin, the locally developed nodular-anhydrite beds within the Salina A-2 carbonate probably were derived from the basinward progradation of sabkha sequences. The Consumers Gas 13009 core (Fig. 8) contains an example of the nodular-anhydrite lithofacies that probably was deposited in a prograding, coastal sabkha sequence. In the basal dolomudstone lying below this anhydrite lithofacies, lenticular anhydrite crystals are pseudomorphous after gypsum. Similar lenticular anhydrite crystals are common in the intertidal deposits of the Trucial Coast (Kinsman, 1969) and were reported in the intertidal sediments of the Gladstone area of Shark Bay, Australia (Davies, 1970). In both areas these gypsum crystals are in oxidized sediments beneath algal mats and crusts. The mats, which are indurated, commonly are broken during storms and make up deposits of flat-pebble conglomerates. These intertidal algal mats and flat-pebble conglomerates are modern analogs of the stromatolite unit and flat-pebble conglomerate overlying the basal dolomudstone in the Consumers Gas 13009 core. The supratidal nodular anhydrite lies above the flat-pebble conglomerate. Hence, the vertical sequence of depositional environments and local extent of the anhydrite indicate that the nodular anhydrite was formed in a basinward-prograding environment similar to that of the Trucial Coast.

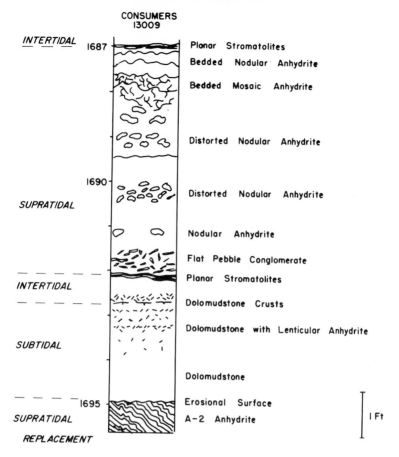

FIG. 8—Description of a core of the basal part, A-2 carbonate unit, showing a prograding-sabkha sequence from the southern Ontario marginal platform.

The regionally extensive anhydrite beds, such as the A-2 anhydrite developed in the upper zone of the A-1 carbonate unit, also were deposited under supratidal conditions. However, it is considered that interstitial fluids were concentrated during sea-level lowering brought about by evaporative drawdown, rather than having been concentrated by basinward progradation of peritidal deposits. In areas where these subaerially exposed surfaces were covered by the A-2 salt, the nodular-anhydrite lithofacies in the upper zone of the A-1 carbonate is thin and poorly developed. The displacive nodular forms—in addition to the overlying salt-displaced carbonate crusts, desiccation cracks, and erosional surfaces—support the idea that the anhydrite was deposited under supratidal or sabkha surfaces. The subaerially exposed areas that were not overlapped by a salt unit also contain anhydrite, but the anhydrite is more massive in these areas because the supratidal environment existed for a longer time. Furthermore, the nodular-anhydrite lithofacies in these areas contains textures that suggest replacement of a carbonate rock. In many places the carbonate-rock textures are still visible in massive anhydrite. Both anhydrite and halite are seen to have replaced carbonate laminations of stromatolites within the A-1 salt, A-1 carbonate, and A-2 salt. Examination of anhydrite replacement in the nodular-anhydrite lithofacies of the lower Salina Group suggests that most replacement took place in a supratidal environment. Evidence of anhydrite replacement of aragonite in the Trucial Coast sabkha (Butler, 1970) bears out the contention that this process can occur in a supratidal environment.

Lithofacies IV: Salt

Description

The salt lithofacies is present in basinal deposits of each of the evaporite units of the lower Salina Group. By volume, salt is the dominant rock type of the basinal deposits. Scarcity of cores and extensive recrystallization of salts prohibit detailed comparison of lateral facies within the A-2 and B salt units. However, the salt facies of the A-1 evaporite unit has been examined in cores from basin-center and marginal areas. Physical characteristics of the salt lithofacies in the basin-center area differ greatly from those of this lithofacies on the interpinnacle shelf in northwestern Michigan.

The salt lithofacies in the A-1 evaporite unit of the basin-center area was described by Dellwig (1953b). His study of the A-1 salt was focused on the core of the Sun 4 Bradley from Newaygo County, Michigan (Fig. 4). Dellwig (1953b) perceived two contrasting types of salt beds within the A-1 salt. The first type is evenly bedded salt (Fig. 9), which also has been seen in marginal deposits of the A-1 salt, in cores of the A-2 and B salt units, and in the A-2 salt in the Goderich salt mine (Fig. 4). Salt beds range from 3 to 8 cm (1.2 to 3.2 in.) thick and alternate with tissue-thin laminae generally composed of anhydrite and(or) carbonate, and trace amounts of pyrite. Average length of halite crystals within this salt is slightly less than 1 cm (0.39 in.).

A crude layering of clear and cloudy salt was observed within this type of salt bed. Cloudiness is due to brine-inclusion-zoned crystals of halite. The fluid inclusions occur within small cubic cavities in the NaCl lattice, where no crystal growth took place during initial crystallization. These brine inclusions generally are considered as evidence of the primary origin of halite crystals (Dellwig, 1953a, 1953b; Wardlaw and Schwerdtner, 1966; Shlichta, 1968; Holdoway, 1974).

Thin sections of cloudy salt have revealed two types of zonation: a cubic-shaped and a chevron-shaped fabric (Fig. 10). The apex of the chevron fabric points upward, normal to bedding. This fabric is predominant in thinly bedded salt on the northwestern interpinnacle shelf.

Dellwig and Briggs (1952) concluded that chevron fabrics commonly observed in thin sections merely are results of slicing through the corners or edges of coronet-shaped, brine-inclusion-zoned halite crystals. However, some geologists (Gottesmann, 1963; Wardlaw and Schwerdtner, 1966) have interpreted chevron fabrics in other evaporite deposits as being formed from upward competitive growth of crystals on the sea floor. Laboratory experiments have demonstrated that chevron-zoned crystals are bottom precipitates (Wardlaw and Schwerdtner, 1966; Arthurton, 1973). The bottom growth of chevron-zoned halite crystals also has been observed in artificial salt pans (Shlichta, 1968; Strakhov, 1970).

Slide-mounted halite slabs, as thick as 15 mm (0.59 in.), were ground and examined at 1 mm intervals for observation of the three-dimensional relationship of brine-inclusion-zoned halite fabrics. Halite crystals in thin-bedded salts of the northwestern interpinnacle area have either chevron-zoned or cubic-zoned fabrics. None of the fabrics was caused by an oblique cut through a coronet-shaped zoned fabric. However, coronet-shaped, inclusion-zoned fabrics were observed in the thicker-bedded salts of the basin-center area. Oblique slicing of these probably could result in a chevron structure, as described by Dellwig and Briggs (1952).

The second type of salt bed described by Dellwig (1953b) is an irregularly bedded, coarsely crystalline salt found only in the basal part of the A-1 salt unit in the central part of the basin. Halite crystals generally are well-developed cubes that are 1 cm (0.39 in.) long, on the average, and that rarely are longer than 3 cm (1.2 in.) (Fig. 11). Almost all the crystals are clear (Dellwig, 1953b). These crystals appear to be primary precipitates that commonly lack the brine-inclusion zonation of the evenly bedded salt layers. However, Dellwig (1953b) attributed the depositional characteristics of this salt bed to recrystallization of the evenly bedded salt. He arrived at an incorrect

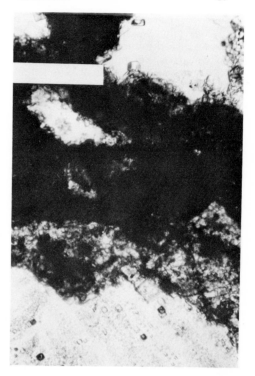

FIG. 9—Core slab of the thinly bedded A-1 salt unit from the northwestern interpinnacle-shelf area. Cloudy zones are due to brine inclusions. Note the very irregular surfaces above two of the cloudy zones. These irregular surfaces generally are overlain by gypsum sands replaced totally or partially by halite. Scale in centimeters.

FIG. 10—Photomicrograph of a thin section, in plane-polarized light, showing the contact of the A-1 carbonate unit (above) and A-1 salt lithofacies (below) from the Sun 4 Bradley, Newaygo County, Michigan (Fig. 4). A halite crystal with inclusions in a distinct chevron zonation is truncated by the laminated, possibly stromatolitic, basal A-1 carbonate unit. Scale bar is 0.5 mm.

assumption, probably because his observations were limited to a single core from the basin-center, in which the basal beds were somewhat recrystallized.

This irregularly bedded salt is interbedded with stringers of the laminated limestone-anhydrite lithofacies. The interbedded salt layers generally contain well-developed, unrecrystallized cubes of halite. Also, sizes of cubes increase vertically within the thicker salt beds (Fig. 11). Within these upper salt stringers, many of the uppermost halite cubes have rounded upper surfaces. These interbedded salts seem to have been precipitates at the sediment-water interface. Along the upper surfaces of these salt beds, areas between cubes are filled by draped laminites that apparently have filled intercrystalline voids left after precipitation of the salt layer. Individual cubes of halite may occur within the laminite stringers. Some of these cubes may have grown within the laminite stringers. Some of these cubes may have grown within the laminated carbonate; there is evidence that the halite may have replaced surrounding laminites.

The salt lithofacies on the northwestern shelf was examined in greater detail than that in the basin-center area. On the northwestern shelf, halite in the upper part of the A-1 evaporite unit consists of thin-bedded salt approximately 2 cm (0.79 in.) thick. Halite crystals within the salt beds are 2 mm long, on the average, but some crystals are longer than 1 cm (0.39 in.). A 10.5-in. (26.7 cm) core slab of the salt lithofacies in the Pan American 1-19 State Kalkaska was thin-sectioned (Fig. 9). The core contains thinly bedded salt that ranges from 2 to 30 mm (0.08 to 1.2 in.) thick. The salt beds are separated by laminae of anhydrite and(or) micrite. Brine-inclusion-zoned halite crystals are the dominant type of halite. The inclusions are in zones parallel to faces of halite cubes, and most commonly form a chevron-like fabric. A few cubic-shaped inclusion fabrics are present. The halite crystals tend to be elongated normal to bedding, and the chevron structure points upward. Many inclusion-free halite crystals are adjacent to brine-inclusion-zoned halite. Less commonly, unzoned halite containing disseminated reddish material is present between zoned crystals. The reddish material probably is a ferrous iron compound such as hematite.

Four of the 20 beds of inclusion-zoned halite in the core slab have very irregular upper surfaces. In each of these four beds, the irregular surface is overlain by rounded grains of halite, each of which has a high content of disseminated, finely crystalline anhydrite (Fig. 12). In some laminae, grains composed almost totally of anhydrite are adjacent to grains having very little anhydrite. The latter grains seem to be the result of replacement by halite. Originally, the grains most probably were gypsum. Diameters of the grains in thin section range from 0.05 to 0.5 mm; however, in individual laminae, the grains are fairly well sorted. An unzoned halite layer (50 m thick) near the

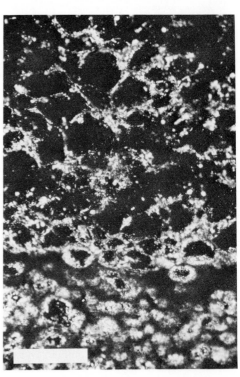

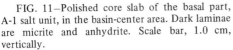

FIG. 11—Polished core slab of the basal part, A-1 salt unit, in the basin-center area. Dark laminae are micrite and anhydrite. Scale bar, 1.0 cm, vertically.

FIG. 12—Photomicrograph of a thin section of gypsum sands, under crossed nicols. The lower one-third of the photomicrograph shows grains replaced by finely crystalline anhydrite and surrounded by micrite. The central black areas of grains are halite. In the upper two-thirds of the photomicrograph, grains are replaced totally by halite. The light-colored, finely crystalline material surrounding grains is anhydrite. Scale bar is 0.5 mm.

middle of the core contains very fine, disseminated red material which presumably is ferric iron. The red material is concentrated around peripheral areas of halite crystals. This halite layer also is overlain by grains that have been replaced by halite. Unfortunately, there is no core available that includes the transition from large cubic halite crystals in the basal part to the overlying evenly bedded salt that contains smaller halite crystals. On the upper basinal slopes halite overlying the laminated limestone-anhydrite lithofacies and coeval to the basin-center salt lithofacies is cubic in habit, although it generally contains brine-inclusion zonations. In the interpinnacle-shelf area in northwestern Michigan, salt overlying the laminated limestone-anhydrite lithofacies is recrystallized, but thin laminae are visible.

Environmental Interpretation

In the basin-center area, basal beds of the salt lithofacies in the A-1 evaporite unit appear to have been precipitated in a deep, hypersaline brine. Fictive-depth analysis, using calculations based on the bromide profile, indicates a depth of 50 m (164 ft) for deposition of the basal salt (Holser, 1966a). However, regional cross sections, which do not take into account major subsidence, suggest somewhat greater depths. Textures, fabrics, and crystal sizes of basal salts in the basin-center area vary greatly from those of basal salts on the shelf and slopes, thus suggesting that the salts precipitated in different environments. Large, well-developed, unzoned cubic halite crystals are in basal beds of the A-1 salt in the basin-center area (Fig. 11). However, they are conspicuously absent in marginal deposits of the A-1 salt unit and in the A-2 and B salt deposits. Absence of these crystals in the shallow-water deposits supports the contention that basal salts in the basin-center area are deeper-water precipitates. The basal basin-center salts are interbedded with stringers of the laminated limestone-anhydrite lithofacies described above as having been deposited in a basinal anaerobic environment. The odor of H_2S emitted in slabbing of these basinal salts, along with the interbedded relation of the salts with deep-water laminites, also corroborate a deeper-water anaerobic environment of deposition.

In basal beds of the A-1 evaporite unit in the basin center, the large cubic halite crystals, which lack brine-inclusion zonation, are suggestive of slow crystallization (Shlichta, 1968). Arthurton (1973) demonstrated through laboratory-controlled crystallization of halite that the deeper the brine, the larger and fewer the halite crystals. Halite is precipitating in various shallow-water hypersaline environments throughout the world, but only one reference to large halite cubes has been found. Cubes as large as 3 cm (1.2 in.) across have been recovered recently at a depth of 34.5 m (113 ft) from the bottom of hypersaline Lake Bonney, Antarctica (Craig et al, 1974). The uncommon depositional environment of these salts, however, may discount the significance of depth to environmental analysis of the Upper Silurian evaporites of the Michigan basin.

The geochemistry or physicochemical conditions of the early Cayugan brine may have resulted in precipitation of large, clear halite cubes in the basal part of the A-1 evaporite unit. This is suggested by the studies of Valyashko (reported in Strakhov, 1970), who maintained that brine-inclusion zonation of halite crystals is derived from varying physicochemical conditions. Physicochemical conditions of a shallow body of brine probably would have varied daily.

Petrographically, the thin-bedded salt deposited on the northwestern interpinnacle shelf is identical to salt of the modern Salina Ometepec described by Shearman (1970) and of the artificial salt pans described by Valyashko (reported in Strakhov, 1970). These Holocene deposits are unlike the upper A-1 salt unit only in presence of voids between brine-inclusion-zoned halite crystals. However, these intercrystalline voids in the modern deposits appear to be analogous to the clear halite crystals adjacent to zoned halite crystals in the A-1 evaporite unit. In the A-1 evaporite, the irregular surfaces that partially truncate inclusion-zoned halite beds suggest periods of dissolution. Valyashko (in Strakhov, 1970) asserted that the "unequal supply of material, caused by changes in evaporation rate and in temperature from day to night, creates a zonal structure in the halite crystal, marked by variable quantities of inclusions. More inclusions are incorporated during more rapid growth than during slow growth." The granular gypsum replaced by halite and anhydrite resembles the interbedded gypsum sands of the Salina Ometepec salt deposits. The close similarity between shallow-water deposits of the Salina Ometepec and thin-bedded salt of the A-1 salt unit strongly invokes a similar shallow-water interpretation for salt deposits on the northwestern interpinnacle-shelf area. Within this salt, LLH-type stromatolites (laterally linked hemispheroids; Logan et al, 1964) stromatolites (Logan et al, 1964) that have been replaced by halite also support the interpretation of a shallow-water environment of deposition. Interfingering of zones of sylvinite and presence of sylvite-filled areas adjacent to chevron-zoned halite crystals suggest periodic desiccation of these salt flats.

Lithofacies V: Sylvinite

Description

The sylvinite lithofacies is composed predominantly of halite and interbedded sylvinite, which is a mixture of sylvite (KCl) and halite. The lithofacies is best developed in the central region of the Michigan basin (Fig. 2). Primarily on the basis of gamma-ray log interpretations, Matthews and Egleson (1974) calculated the area covered by this lithofacies as 13,000 sq mi (33,700 sq km). In the basin-center region, very few wells have been drilled into the sylvinite lithofacies, which in Ogemaw County (Fig. 4) is 8,000 ft (2,400 m) deep. Sylvinite has been identified in the Dow 8 Chemical, Midland County (Anderson and Egleson, 1974); Pan American 1-19 State Kalkaska, Kalkaska County (Huh, 1974); Pan American 1-14, State Union, Grand Traverse County (Mesolella et al, 1974); and Amoco 1 Hunt Club, Ogemaw County (Nurmi, 1974) (for locations of wells, see Fig. 4). In cores from Kalkaska County on the northwestern shelf, the sylvinite lithofacies is poorly developed and contains several thin sylvinite zones. In the central part of the basin, however, the lithofacies is approximately 100 ft (30.5 m) thick. In the northern part of the basin-center region, the lithofacies lies at the top of the A-1 salt unit, whereas about 40 mi (64 km) southward, in Midland County (Fig. 4), the lithofacies is in the middle of the A-1 salt unit (Fig. 6).

Description of cores in the sylvinite lithofacies in the basin-center region is limited herein to the Amoco 1 Hunt Club core from Ogemaw County (Fig. 4). It includes the upper 57.4 ft (17.5 m) of a probable 98 ft (30 m) of the sylvinite lithofacies. Halite in the Amoco core is highly recrystallized and commonly is crystallographically continuous across the core. The halite is massive throughout the upper half of the core but is progressively bedded in the lower half. The lowest 2 ft (0.61 m) are thinly laminated but highly recrystallized. Beds of sylvinite are throughout the core and they range from less than 1 cm (0.39 in.) to 14 cm (5.5 in.) thick. The sylvite is reddish white with the exception of a zone from 7,892.8 to 7,893 ft (2,405.7 to 2,406 m) where the sylvite is milky white. Reddish coloration of sylvite has been shown to be due to hematite or goethite (Wardlaw, 1968; Adams, 1969; Braitsch, 1971), but the significance of these minerals to the depositional environment or diagenesis of potassium salts is undetermined. The halite-sylvite mineral boundaries are concave. Small sylvite grains (less than 0.2 cm [0.08 in.]) commonly are almost spherical; larger grains are amoeboid and generally are elongated horizontally.

White, opaque calcium-borate nodules ranging in diameter from less than 1 mm to 30 mm have been noted throughout the Amoco 1 Hunt Club core from the northern basin-center region. Large nodules rarely are attached, but may appear to be a compact aggregate of small nodules. In this core, at the depth of 7,907 ft (2,410 m), there is part of a possible solution (depositional or diagenetic) depression, 10 cm (3.9 in.) deep, containing nodules and wisps of calcium borate. Although the calcium-borate nodules are associated with sylvinite beds, they also are in halite beds. No borate nodules have been found in the Dow 8 Chemical core (R. D. Matthews, personal communication, 1974).

The X-ray diffraction patterns of five nodules taken at approximately 5-ft (1.5 m) intervals in the core, starting at 7,900 ft (2,408 m) deep, were virtually the same. However, the X-ray diffraction peaks have not been matched with a mineral in the

ASTM file. Qualitative emission-spectroscopic analysis of a nodule that had been submerged in distilled water for 2 hrs revealed calcium, boron, and sodium. Quantitative analysis of a nodule by atomic absorption spectroscopy at the Texaco Research Laboratory at Beacon, New York, showed a chemical composition of 21.4% Ca, 18.2% B, 1.6% Na, and 0.06% K. The nodule was 14% soluble in hot water. Qualitative X-ray fluorescence analysis of the insoluble residue showed no sulfur but did reveal chlorine, indicating that the mineral has chloride attached.

Detailed petrographic analysis of the basin-center lithofacies has not been undertaken in this study or by Dow Chemical Co. (R. D. Matthews, personal communication, 1974). Analysis of the complex paragenesis of late-stage evaporite minerals is beyond the scope of this investigation. X-ray diffraction analysis of the Amoco 1 Hunt Club core at 1-ft (0.3 m) intervals, and at zones where leaching indicated presence of a mineral more soluble than halite, failed to reveal any minerals other than halite, sylvite, or calcium borate.

Chemical analyses of the basin-center sylvinite lithofacies in the Dow 8 Chemical core from a well 40 mi (64 km) south of the Amoco 1 Hunt Club well have been recorded by Anderson and Egleson (1970) of Dow Chemical Company. They report that the sylvinite lithofacies is "exceptionally clean" in the core. No minerals other than halite and sylvite have been identified (R. D. Matthews, personal communication, 1974). Whole-rock analysis has shown that the magnesium content of the sylvite beds ranges from 21 to 71 ppm. Throughout the sylvinite lithofacies in this core, concentrations of sulfate and magnesium are in trace amounts, thus explaining the absence of other common late-stage evaporite minerals.

Anderson and Egleson (1970) established that sections high in sylvinite correlate with high API-unit readings on gamma-ray logs. This allows correlation of the sylvinite lithofacies in the basin-center region. In basin-margin areas, however, sylvinite beds generally are thinner than 1 cm (0.39 in.) thick; only core analysis can determine unequivocally the presence of a potassium salt in these areas.

Examination of core slabs from the Pan American 1-14 State Union and the Pan American 1-19 State Kalkaska has provided insight to the sylvinite lithofacies of the upper half of the A-1 evaporite on the northwestern shelf. Sylvite in these cores predominantly is reddish white, presumably because of hematite or goethite. Rare blebs of sylvite are milky white. The sylvite is developed irregularly between laminae of halite but is nowhere continuous across the width of a core. Sylvite zones 2 to 5 mm thick commonly overlie laminae of primary, inclusion-zoned halite crystals (Fig. 13). A few blebs of sylvite 1 to 2 mm long lie between halite crystals containing chevron-shaped, inclusion-zoned fabrics. Within the halite laminae overlying the sylvite zones, inclusion-zoned halite crystals are rare. Two small white nodules with a high content of boron and calcium were found within anhydrite laminae in this facies on the northwestern shelf.

Environmental Interpretation

The extreme concentration of sea water necessary to precipitate sylvite and calcium-borate nodules requires that the basinal hypersaline sea was almost totally evaporated during deposition of the basin-center sylvinite lithofacies. Schmalz (1969) showed clearly that bedded potassium salts can accumulate in a deep basin only if the sea desiccates. He demonstrated that if the basin remained full of water, as has been proposed by Matthews (1970) and Matthews and Egleson (1974), the entire basin would have been filled by sulfates and halite before the sea became concentrated enough for precipitation of potassium salts.

Distribution of the sylvinite lithofacies at the top of the A-1 evaporite unit in the basin-center area appears to have been restricted to the northern side of the basinal depression (Fig. 6). This can be explained by lateral segregation of the brine. Evidence of channels connecting the Michigan basin, the southern Illinois basin (Shaver, 1974), and the Ohio sub-basin during the early Cayugan could suggest that waters came from the south, across the subaerially exposed southern shelf. The low content of sulfate and magnesium in the basin-center sylvinite lithofacies (Anderson and Egleson, 1970)

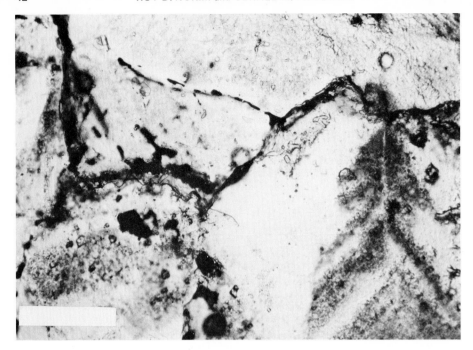

FIG. 13–Photomicrograph of a thin section, plane-polarized light, showing sylvite above and between brine-inclusion-zoned halite crystals. Sample of the sylvinite lithofacies from the northwestern interpinnacle-shelf area. Scale bar is 0.5 mm.

thus may be explained by dolomitization and by precipitation of gypsum interstitially on the southern shelf. Loss of magnesium through dolomitization of shelf sediments also would account for absence of magnesium-bearing evaporite minerals and presence of calcium-borate nodules in this lithofacies. The dominant paleowind direction at this time may have been to the northwest (Stuart, 1971); it would have accentuated the salinity gradient to the north. The lateral salinity gradient could have resulted in lateral segregation of evaporite-mineral precipitation in a fashion described by Scruton (1953) or in a facies change in a marine-playa setting.

Thin sections of the sylvinite lithofacies on the northwestern shelf reveal that much of the halite associated with sylvite is chevron-zoned and is identical petrographically to halite in the salt facies of the northwestern shelf. Sylvite that commonly is adjacent to inclusion-zoned halite crystals may be interpreted as having crystallized later than the zoned halite. Bromide analysis of interbedded sylvite and halite in the basin-center sylvinite lithofacies indicates that the minerals crystallized at different concentrations (Matthews and Egleson, 1974). Abrupt changes in brine concentration are consistent with deposition in very shallow water.

A modern environment somewhat analogous to environment of deposition of the sylvinite lithofacies on the northwestern interpinnacle shelf is the Salina Ometepec on the Colorado River delta, described by Shearman (1970). However, waters entering the Ometepec salina are of normal-marine salinities, whereas waters of the Michigan basin would have been hypersaline during deposition of the sylvinite lithofacies. Petrography of the Salina Ometepec halite is very similar to that of halite in the sylvinite lithofacies, except that intercrystalline voids in halite of the Salina Ometepec are not filled. Intercrystalline brines of the Salina Ometepec, analyzed by Smith (1971), contain notably high concentrations of potassium and magnesium. If these brines were not diluted by ground water or by normal-marine water that flooded salt flats during storms, later-stage evaporite minerals would be found in voids and dissolution channels around the zoned halite crystals. High concentrations of potassium and magnesium were also found in intercrystalline brines of the salt flats of Laguna Ojode Liebre, Baja California, by Holser (1966b) and Phleger (1969). Holser (1966b) reported early diagenetic polyhalite $(K_2 MgCa_2 (SO_4)_4 H_2 O)$ that had replaced gypsum crystals in tidal-flat sediments of this lagoon. On the western margin of the central part of the Michigan basin, early diagenetic replacement of gypsum crystals probably had also occurred in upper beds of the A-1 salt. Lucas (1954) identified carnallite $(KMgCl_3 \cdot 6H_2 O)$ pseudomorphous after gypsum in a core from this area. In the same core, polyhalite that may have replaced anhydrite was identified by Dellwig (1953b).

Studies of evaporite deposits in the Soviet Union by Valyashko (1972) also substantiate that the basin-center sylvinite lithofacies was the result of near-desiccation of the basin. Valyashko (1972) concluded that primary sylvite is precipitated with halite only in a "playa-lake" stage. From study of modern and ancient evaporite deposits Valyashko (1972, p. 49) affirmed that "a marine salt-forming basin will reach the stage of a playa lake at some time before the precipitation of the potassium salts." In continental playas, potassium minerals precipitate interstitially between previously precipitated halite crystals.

Calcium-borate nodules in the sylvinite lithofacies demonstrate that the basin-center brine was so concentrated that the sea must have been totally desiccated, at least periodically. In fact, calcium-borate nodules formerly were thought to be diagnostic of evaporite deposition in terrestrial salt playas. However, Braitsch (1971) stated that these nodules are much too rare to be used as environmental index minerals. Nikolaev and Chelishcheva (1940) determined experimentally that by evaporation, a brine would be concentrated to levels greater than 1.4% boron in order to precipitate borates. This degree of concentration requires a desiccated basin, not a brine-filled basin as proposed by Matthews (1970).

Lithofacies VI: Nonplanar Stromatolite

Description

The nonplanar-stromatolite lithofacies commonly is closely associated with the nodular-anhydrite lithofacies and its stratigraphic distribution is similar. Within the basinal depression, these nonplanar stromatolites are associated with the nodular-anhydrite lithofacies or are interbedded within the salt units. Algal growth structures are highly distorted and barely recognizable within the early diagenetic nodular-anhydrite lithofacies. However, directly below the salt beds, growth structures generally are preserved either by halite replacement of organic matter and(or) by early filling of porous algal structures by halite.

The nonplanar-stromatolite lithofacies also is common within the A-1 carbonate unit deposited above the eroded Niagaran pinnacle reefs and on the marginal platforms. Felber (1964) was first to recognize the extensive development of stromatolites in the A-1 carbonate unit, above Niagaran pinnacle reefs in southeastern Michigan. The occurrence and growth forms of stromatolites overlying the Niagaran Belle River Mills pinnacle reef in southeastern Michigan were described by Gill (1973).

In cores of the lower Salina Group, the most common nonplanar stromatolite is the laterally linked (LLH) form. These hemispheroids generally are of low amplitude and have basal domal radii that are as large as 3 cm (1.2 in.). Relief of individual laminae rarely exceeds 1 cm (0.39 in.). Central parts of the hemispheroids commonly contain irregularly oriented laminae and greater amounts of replacement and(or) infilling halite.

Vertically stacked stromatolites (SH type) are present in an anhydrite-carbonate stringer in the upper part of the A-2 salt unit in the basin-center area. These stromatolites generally are less than 2 cm (0.79 in.) high and range in width from 1 to 4 cm (0.39 to 1.6 in.). The growth form is distorted by anhydrite; nevertheless, a few forms are preserved by partial replacement and(or) infilling by halite. Similar SH-type stromatolites, 2 cm (0.79 in.) high, are developed in the transitional zone between the A-2 (anhydrite) evaporite and the A-1 carbonate lying above a Niagaran pinnacle reef in northwestern Michigan (Huh, 1973).

Oncolites (SS-type stromatolites) are present in channel carbonate sands at the top of the A-1 carbonate in the Goderich salt mine, Huron County, Ontario (Fig. 4). Fragmented stromatolite clasts also are contained within this detrital sand. The oval oncolite structures are as large as 1.5 cm by 2.5 cm (0.6 in. by 1 in.) and have a high concentration of replacement and(or) infilling halite.

Environmental Interpretation

It generally is now accepted that stromatolite forms are related to environmental conditions (Hoffman et al, 1968; Ginsburg, 1971). The only oncolites found in the Michigan basin, other than in salt beds, were those observed within channel-like erosional features underlying the A-2 salt in the Goderich mine. These channel-like features, along with fragments of mats and carbonate sand-sized particles also associated with the oncolites, suggest that this stromatolite form was developed in shallow, agitated water.

The nonplanar stromatolites discussed herein are shallow-water forms. The environmental interpretation of each of these forms in the Michigan basin generally is consistent with interpretations set forth by Logan et al (1964). Thus, nonplanar stromatolites are valuable indicators of Cayugan paleoenvironments. The LLH, SH, and SS stromatolite forms within salt beds of the lower Salina Group, and other sedimentologic features, indicate that these stromatolites were formed in shallow-water environments and on margins of shallow hypersaline seas during flooding of playalike salt flats. Somewhat analogous nonplanar stromatolite structures were described from the margins of salt lakes in Western Australia (Clarke and Teichert, 1946) and from the Great Salt Lake, Utah (Eardley, 1938; Carozzi, 1962).

Lithofacies VII: Stromatolite-Peloid

Description

The stromatolite-peloid lithofacies commonly is developed as thin zones (0.5 to 2.0 ft or 0.15 m to 0.6 m) of interbedded peloids and stromatolites within the A-1 carbonate unit. This lithofacies has been found in the upper beds of many A-1 carbonate cores from pinnacle reefs in southeastern and northwestern Michigan, and southern Ontario. It also is present in interpinnacle areas of both the northwestern and southeastern regions of the basin, and was identified in the upper 20 ft (6 m) of these A-1 carbonate cores from the basin-center area. The stromatolite-peloid lithofacies commonly is dolomitized on the pinnacle reefs and in the basin.

Environmental Interpretation

The interstratified peloids and stromatolites comprising this lithofacies strongly suggest a shallow-water environment of deposition for the lithofacies. The partial dolomitization of alternating laminae is characteristic of an algal origin (Friedman et al, 1973). In addition, the abundant carbonaceous matter within the lithofacies indicates a high original content of organic matter within the peloids and stromatolites. Gill (1973), Huh (1973), and Budros (1974) proposed that these peloids in the A-1 carbonate are of fecal origin. Although fecal pellets in modern carbonate environments generally are not as long as peloids of the Salina Group, fecal pellets 10 to 15 mm (0.39 to 0.6 in.) long are associated with laminated algal sediments in Shark Bay, Australia (Davies, 1970). The elongated shape of the peloids, carbonaceous matter within them, and association of peloids with stromatolites present a cogent argument for the fecal origin of these particles.

An interpretation of shallow-water deposition for this stromatolite-peloid lithofacies clearly is strengthened by the similar interpretation of adjacent lithofacies. The stromatolite-peloid lithofacies is interbedded with sabkha-like nodular and enterolithic anhydrite on the northwestern interpinnacle shelf. It also underlies sabkha-like nodular A-2 anhydrite on the pinnacle reefs. The combination of sabkha evaporites, pellets, and stromatolites is reminiscent of lagoons of the Trucial Coast (Kendall and Skipwith,

1969; Kinsman, 1969; Butler, 1970). Moreover, the position of the stromatolite-peloid lithofacies near the top of the A-1 carbonate unit in the center of the basin is consistent with a postulated evaporative lowering of sea level that culminated in deposition of the overlying A-2 salt unit.

Lithofacies VIII: Peloid-Ooid

Description

The peloid-ooid lithofacies is relatively thin, and is in the basal part of the A-2 carbonate of the basin-center region. This lithofacies was observed in two cores, one from the basin center in Midland County, and one from near the basin-center slope, in Oceana County (Fig. 4). Areal distribution of this lithofacies is poorly defined because of scarcity of cores of the A-2 carbonate unit in the central part of the basin. Examination of bit cuttings of the basal part of the A-2 carbonate unit from other wells in the basin-center region failed to reveal ooids or peloids. Absence of these allochems in cuttings may be due to thinness of the lithofacies in the basal 10 ft (3 m) of the A-2 carbonate unit, rather than to nondeposition of ooids or peloids. However, State of Michigan sample descriptions report ooids throughout the A-2 carbonate in back-platform areas of northern Michigan. Oolitic zones of the A-2 carbonate on the northern platform were mapped by Tremper (1973). They extend from northern Antrim County to northern Presque Isle County (for location of counties, see Fig. 4). On the northern platform, the A-2 carbonate, including oolitic zones, is essentially a dolostone, whereas in the basin-center area it is predominantly limestone, including the peloid-ooid lithofacies.

The lower part of the peloid-ooid lithofacies contains peloidal allochems and clasts of anhydrite, in addition to ooids. Also contained are some mud-supported and clustered (grapestone) ooids. The upper part of this lithofacies contains well-sorted, nonclustered ooids that are predominantly grain-supported. A few ooids examined in the Carter 12 Lauber core from Oceana County (Fig. 4) were partially or totally leached; the oomolds were filled by anhydrite or halite. Diameters of ooids are 0.5 to 0.6 mm, on the average. Lath-shaped anhydrite crystals are present within individual ooids and cross ooid boundaries. In the Dow 8 Chemical core, Midland County (Fig. 4), few individual ooids remain. The oomolds are filled by halite.

Environmental Interpretation

Although the peloid-ooid lithofacies is in the center of the Michigan basin, it was not deposited in deep water. Numerous occurrences of ooids in ancient shallow-water deposits clearly suggest that the ooids are of shallow-water origin only. Ooids are forming today both in shallow, current-swept environments (Eardley, 1938; Newell et al, 1960) and in shallow, relatively undisturbed environments (Rusnak, 1960; Freeman, 1962; Friedman, 1964; Friedman et al, 1973; Loreau and Purser, 1973). Furthermore, these modern ooid forms indicative of low-energy conditions are forming only in restricted sea-marginal lagoons that become hypersaline, at least periodically.

In order to conform to the inferred general paleogeography of the Michigan basin during early Cayugan time, the basinal peloid-ooid lithofacies could not have been deposited in a marginal lagoon. However, the chemical and(or) biochemical conditions requisite for formation of ooids may have existed in a shallow, below-sea-level body of water within the Cayugan basinal depression in Michigan. Although ooids form in shallow-water environments, they are transported into deeper water. Transported ooids were found in the deep-ocean deposits marginal to the Bahama Platform (Friedman, 1964). Ooids also were identified in axial regions of the Persian Gulf (Loreau and Purser, 1973) and in deep waters of the Yucatan Shelf (Logan et al, 1969). In both of these identifications, the present deep-water position of the ooids was explained by deposition during glacial stages of low sea level, which were followed by post-glacial transgressions. It is proposed herein that in the center of the Michigan basin, the peloid-ooid lithofacies was deposited similarly during a low stage of sea-level. However, the low stage in the Michigan basin was due to evaporative drawdown, not to glacier-controlled sea-level fluctuations. Peloids, anhydrite lithoclasts, poor sorting of ooids, and clustered ooids in the lower zones depict a quiet, shallow-water environment of deposition with occasional disturbances of the bottom by waves or currents. The upper zone suggests wave or current activity during formation of ooids, or possibly a re-working of ooids in the lower zone.

The shallow-water interpretation advanced for deposition of the section underlying the basinal peloid-ooid lithofacies also supports the premise that the ooids and peloids were formed within the center of the basin and were not resedimented; this underlying section is composed of interbedded salt and carbonate, and contains peloids and SH-type stromatolites.

Lithofacies IX: Laminated Carbonate Mudstone

Description

Laminated carbonate mudstones are the most common rock type within the A-1 and A-2 carbonate units in the Michigan basin. Lithology of the laminated carbonate mudstone lithofacies is limestone, dolomitic limestone, or dolostone. The quantity of anhydrite, halite, carbonaceous matter, argillaceous sediment, and pyrite within this lithofacies varies from none to abundance. In the basin-center area, this lithofacies is particularly well-laminated. The laminae are dominantly even and parallel, because of the high content of carbonaceous matter and argillaceous sediment. In shalf and plat-form-slope regions this lithofacies generally contains less carbonaceous and argillaceous matter than it does in the central part of the basin. Also in these regions, the laminae generally are subparallel and wavy.

Environmental Interpretation

The laminated carbonate mudstone lithofacies is presumed to have been deposited in subtidal environments varying from just below wave base to deep subtidal. No sedimentologic criteria for subaerial exposure or for wave and(or) current action were present in the hundreds of feet of core of this lithofacies that were examined, except at places where mudstones interfingered with or were adjacent to evaporite strata.

During more arid climatic periods, direct inorganic precipitation and sulfate-reducing bacteria may have been responsible for much of the lime mud deposited in carbonates of the lower Salina Group in the basin-center region. The dearth of fossils and abundance of pyrite in the center of the basin—as well as sulfate rocks in the marginal areas of this lithofacies—all fit the sulfate-replacement mechanism (Friedman, 1972). It is advanced herein that direct inorganic precipitation, organic precipitation (blue-green algae and bacteria), and resedimentation of detrital carbonate mud from exposed limestones on margins of platforms, may have all been causative factors in the accumulation of lime muds of this lithofacies.

Summary of Depositional Environments

Interpretation of the depositional environment of an evaporite lithofacies is complicated by the possibility of repeated fluctuations of sea level. These fluctuations can be very rapid, that is in the context of geologic time, and deep-water deposits may be interbedded with shallow-water deposits. An interpretation is hindered because of lack of a permanent sea-level datum, and also because there are few depth criteria.

The first Cayugan evaporites deposited in the central part of the basin were lenses of gypsum, later altered to anhydrite. In places where there was relatively greater concentration of these crystals, diagenetic alteration, perhaps along with additional precipitation of calcium sulfate, resulted in macroscopic laminae of anhydrite—the laminated limestone-anhydrite lithofacies. These undolomitized laminites, which occur only in the basin-center area, were deposited in the deepest water of the early Cayugan. Exact determination of water depth based on sedimentological evidence is not possible, but the depth probably was more than 50 m (more than 164 ft). The laminated limestone-anhydrite lithofacies also is in the middle of the A-1 carbonate unit in the central part of the basin, and probably is coeval with the Rabbit Ears nodular to nodular-mosaic anhydrites developed around the pinnacle reefs. However, gypsum lenses in

FIG. 14—Oscillation ripples on salt in the ceiling of the Goderich salt mine Huron County, Ontario (Fig. 4). Rippled surfaces on salt were observed at two localities more than 1 mi (1.6 km) apart. For scale, rock bolts (see arrow) are approximately 1 ft (0.3 m) apart.

the A-1 carbonate were replaced by calcite rather than anhydrite as, in the A-1 evaporite. The associated laminae were brecciated chaotically, in a manner suggesting that salt was interbedded with the laminae but was removed by dissolution shortly after deposition. Basal laminites in the A-1 evaporite in the basin-center area were interbedded with salt also, and in many places overlie minor dissolution surfaces of the underlying salt beds. Brecciated laminites were observed in southwestern Michigan, but in this area the brecciation is nonchaotic, indicating presence of a substantial overburden at the time of dissolution. Stratigraphic analysis of salts of the lower Salina Group revealed that dissolution was post-Devonian and resulted in the salt-cored anticlinal traps in southwestern Michigan (Nurmi, 1974).

In the basin-center area, the salt lithofacies is composed of unzoned cubic halite and is interbedded with the laminites. Like the laminites, the halite is interpreted as having been deposited in a deeper-water environment. The rare coronet-inclusion fabrics within halite crystals may be floundered hoppers that crystallized at the brine-air interface, as suggested by Dellwig (1953a). However, in the same core studied by Dellwig (1953a) chevron-zoned halite (Fig. 10) was identified in the salt lithofacies at the top of the A-1 evaporite, indicating progressive shallowing of the brine.

A well-developed lithofacies of grasslike gypsum molds and laminated anhydrite has been recognized thus far only on the northwestern interpinnacle shelf. This lithofacies probably was widespread on the margins of the basin, but conversion of gypsum to anhydrite has obscured the original crystal forms. Although knowledge of distribution of this lithofacies is limited, the lateral and vertical facies indicate shallower deposition of this lithofacies than of the underlying lithofacies of interbedded salt and laminite. Similar vertically oriented crystals of gypsum replaced by halite are associated with stringers within the middle of the A-2 salt in southwestern Ontario. An interpretation of shallow-water deposition is supported by dissolution channels, nonplanar

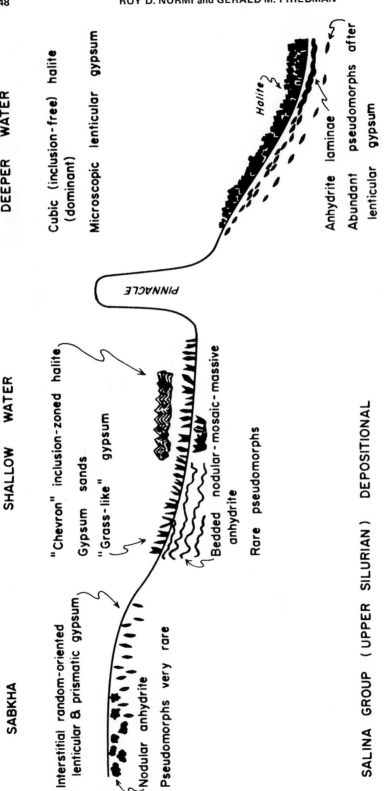

FIG. 15—Idealized, schematic cross section showing the northwestern part of the Michigan basin during the deposition of the lower half, A-1 evaporite, before near-desiccation and precipitation of the sylvinite lithofacies.

stromatolites, possible desiccation cracks, oncolites, nodular anhydrite, and oscillation ripples (Fig. 14) associated with the stringers.

Chevron-zoned halite of the salt lithofacies overlies the grasslike gypsum-molds and laminated-anhydrite lithofacies on the northwestern interpinnacle shelf, and is interpreted as having been deposited in even shallower water than the underlying rock. In this area, the salt lithofacies is interbedded with the sylvinite lithofacies, and with gypsum sands that have been replaced by anhydrite and halite. Because an unrecrystallized core of the entire A-1 evaporite from the central part of the basin is not available, it is impossible to determine how much of the salt in this region was deposited in shallow water. However, chevron-zoned halite at the top of the A-1 salt in the basin-center area supports the contention that water in the center of the basin became progressively shallower during deposition of the A-1 evaporite. Generalized depositional environments and diagenetic facies of the northwestern Michigan basin are shown diagrammatically in Figure 15.

The sylvinite lithofacies interbedded in the salt lithofacies is thought to have precipitated in the shallowest water of the A-1 evaporite unit in the Michigan basin. Sylvite was precipitated in intercrystalline voids of chevron-zoned halite. Sylvite and tiny borate nodules within the sylvinite lithofacies on the northwestern interpinnacle shelf may have precipitated on broad, marginal salt flats during periods of low sea level. However, sylvinite in the center of the basin would have been deposited in a marine-playa stage of an isolated marine basin.

Conclusion

Amadeus Grabau is remembered as an advocate of a desert origin of Salinan evaporites. However, more pertinent to this study are his conclusions that during the Middle-Late Silurian, the Michigan basin was the deepest basin in the northeastern United States, and that withdrawal, presumably by evaporation, of the Niagaran sea from the Michigan basin resulted in precipitation of the basal salt of the Salina Group. More than 50 yrs before discovery of potash salts in the basal Salinan salt, Grabau deduced that potash salts would be associated with this lowermost unit. The interpretation of depositional conditions described in this paper is similar to the model invoked by Grabau to explain evaporites in the Michigan basin (Grabau and Sherzer, 1910; Grabau, 1913). Nevertheless, in study of basin-center evaporites, any of the proposed stratigraphic geochemical models may be partially or totally applicable. Only through an understanding of the sedimentology and stratigraphy of the evaporite lithofacies can a valid interpretation of depositional history be reached. Although no apparent modern analogs of a basin-center evaporite exist, the rocks themselves are the key to the past.

References Cited

Adams, S. S., 1969, Bromine in the Salado Formation, Carlsbad potash district, New Mexico: New Mexico State Bureau Mines and Mineral Resources Bull. 93, 122 p.

Anderson, R. J., and G. C. Egleson, 1970, Discovery of potash in the A-1 Salina salt in Michigan, *in* W. A. Kneller, ed., Proc. 6th ann. forum on geol. of industrial minerals: Mich. Geol. Survey, Misc. 1, p. 15-19.

Anderson, R. Y., et al, 1972, Permian Castile varved evaporite sequence, West Texas and New Mexico: Geol. Soc. America Bull., v. 83, p. 59-86.

Arthurton, R. S., 1973, Experimentally produced halite compared with Triassic layered halite-rock from Cheshire, England: Sedimentology, v. 20, p. 145-160.

Berner, R. A., 1971, Principles of chemical sedimentology: New York, McGraw-Hill, 240 p.

Bischof, F., 1875, Die Steinsalzwerke bei Stassfurt: Halle, Pfeffer, 2nd ed., 95 p.

Braitsch, O., 1971, Salt deposits, their origin and composition, translation by P. J. Burek, and A. E. M. Nairn: Berlin, Springer-Verlag, 297 p.

Bramkamp, R. A., and R. W. Powers, 1955, Two Persian Gulf lagoons: Jour. Sed. Petrology, v. 25, p. 139-140.

Branson, E. B., 1915, Origin of thick gypsum and salt deposits: Geol. Soc. America Bull., v. 26, p. 231-242.

Budros, R., 1974, The stratigraphy and petrogenesis of the Ruff formation, Salina Group in southeast Michigan: Unpub. M.S. thesis, Univ. Michigan, 178 p.

Bush, P. R., 1973, Some aspects of the diagenetic history of the sabkha in Abu Dhabi, Persian Gulf, *in* B. H. Purser, ed., The Persian Gulf: New York, Springer-Verlag, p. 395-407.

Butler, F. P., 1969, Modern evaporite deposition and geochemistry of coexisting brines, the sabkha, Trucial Coast, Arabian Gulf: Jour. Sed. Petrology, v. 39, p. 70-89.

―― 1970, Holocene gypsum and anhydrite of the Abu Dhabi sabkha. Trucial Coast. An alternative explanation of origin, Third Symposium on Salt: Northern Ohio Geol. Soc., v. 1 p. 120-152.

Carozzi, A. V., 1962, Observations on algal biostromes in Great Salt Lake, Utah: Jour. Geology, v. 70, p. 246-252.

Clarke, F. C., and C. Teichert, 1946, Algal structures in a Western Australia salt lake: Am. Jour. Sci., v. 244, p. 271-276.

Cole, L. H., 1915, The salt deposits of Canada and the salt industry: Canada Dept. Mines Rept. No. 325, 152 p.

Cook, C. W., 1914, The brine and salt deposits of Michigan: their origin, distribution, and exploitation: Mich. Biol. Geol. Survey Pub. 15, Geol. Ser. 12, 188 p.

Craig, J. R., R. D. Fortner, and B. L. Weand, 1974, Halite and hydrohalite from Lake Bonney, Taylor Valley, Antarctica: Geology, v. 2, p. 389-390.

Davies, G. R., 1970, Algal-laminated sediments, Gladstone Embayment, Shark Bay, Western Australia, *in* B. W. Logan, et al, eds., Carbonate sedimentation and environments, Shark Bay, Western Australia: AAPG Mem. 13, p. 169-205.

Dellwig, L. F., 1953a, Origin of the Salina salt of Michigan: Jour. Sed. Petrology, v. 25, p. 83-110.

―― 1953b, Origin of the Salina salt of Michigan: Unpub. Ph.D. thesis, Univ. of Michigan, 98 p.

―― and L. I. Briggs, 1952, Textural relationships in the Salina salt of Michigan (abs.): Geol. Soc. America Bull., v. 63, p. 1242.

Eardley, A. J., 1938, Sediments of Great Salt Lake, Utah: AAPG Bull., v. 22, p. 1305-1411.

Felber, B. E., 1964, Silurian reefs of southeastern Michigan: Unpub. Ph.D. thesis, Northwestern Univ., 124 p.

Freeman, T., 1962, Quiet water oolites from Laguna Madre, Texas: Jour. Sed. Petrology, v. 32, p. 475-483.

Friedman, G. M., 1964, Early diagenesis and lithification of carbonate sediments: Jour. Sed. Petrology, v. 34, p. 643-655.

―― 1972, Significance of Red Sea in problem of evaporites and basinal limestones: AAPG Bull., v. 56, p. 1072-1086.

―― et al, 1973, Algal mats, carbonate laminites, ooids, oncolites and pellets in sea-marginal hypersaline pool, Gulf of Aqaba (Elat), Red Sea: AAPG Bull., v. 57, p. 541-557.

―― and J. E. Sanders, 1967, Origin and occurrence of dolostones, *in* G. V. Chilingar et al, eds., Carbonate rocks: Amsterdam, Elsevier, p. 267-348.

Gill, D., 1973, Stratigraphy, facies, evolution, and diagenesis of productive Niagaran-Guelph reefs and Cayugan sabkha deposits, the Belle River Mills gas field, Michigan basin: Unpub. Ph.D. thesis, Univ. Michigan, 275 p.

Ginsburg, R. N., 1971, Recent stromatolites, *in* R. N. Ginsburg et al, Geology of calcareous algae, notes for a short course: Comparative Sedimentology Laboratory, Univ. Miami, p. 12.1-12.3.

Gottesmann, W., 1963, Eine häufig auftretende Struktur des Halite im Kaliflöz Stassfurt: Geologie, v. 12, p. 576-581.

Grabau, A. W., 1913, The origin of salt deposits with special reference to the Siluric salt deposits of North America (with discussion): Mining Metallurgy Soc. America Bull. 57, v. 6, p. 33-44.

―― and W. H. Sherzer, 1910, The Monroe Formation of southern Michigan and adjoining regions: Mich. Geol. Biol. Survey, Publ. 2 (1909), Geol. Ser. 1, 248 p.

Hardie, L. A., and H. P. Eugster, 1971, The depositional environment of marine evaporites: a case for shallow, clastic accumulation: Sedimentology, v. 16, p. 187-220.

Harrison, A. G., and H. G. Thode, 1958, Mechanism of the bacterial reduction of sulphate from isotope fractionation studies: Trans. Faraday Soc., v. 54, p. 84-92.

Hoffman, P. F., B. W. Logan, and C. D. Gebelein, 1968, Biological versus environmental factors governing the morphology and internal structures of recent algal stromatolites in Shark Bay, Western Australia (abs.): Geol. Soc. America, Abs. with Programs, p. 28-29.

Holdoway, K. A., 1974, Behavior of fluid inclusions in salt during heating and irradiation, *in* A. H. Coogan, ed., Fourth symposium on salt: Northern Ohio Geol. Soc., v. 1, p. 303-312.

Holser, W. T., 1966a, Bromide geochemistry of salt rocks: Northern Ohio Geol. Soc., Second Symposium on Salt, v. 1, p. 248-275.

―― 1966b, Diagenetic polyhalite in Recent salt from Baja California: Amer. Mineralogist, v. 51, p. 99-109.

Huh, J. M. S., 1973, Geology and diagenesis of the Niagaran pinnacle reefs in the northern shelf of the Michigan basin: Unpub. Ph.D. thesis, Univ. Michigan, 253 p.

—— 1974, Core examination in subsurface laboratory *in* R. V. Kesling, ed., Silurian reef-evaporite relationships: Mich. Basin Geol. Soc. Ann. Field Excursion, 1974, p. 79-88.

Jones, C. L., 1965, Petrography of evaporites from the Wellington Formation near Hutchinson, Kansas: U.S. Geol. Survey Bull. 1201-A, 70 p.

Kemp, A. L. W., and H. G. Thode, 1968, The mechanism of the bacterial reduction of sulphate and of sulphite from isotope fractionation studies: Geochim. et Cosmochim. Acta, v. 32, p. 71-93.

Kendall, C. G. St. C., and P. A. D'E. Skipwith, 1969, Holocene shallow-water carbonate and evaporite sediments of Khor al Bazam, Abu Dhabi, southwest Persian Gulf: AAPG Bull., v. 53, p. 841-869.

King, R. H., 1947, Sedimentation in Permian Castile sea: AAPG Bull., v. 31, p. 470-477.

Kinsman, D. J. J., 1969, Modes of formation, sedimentary associations and diagnostic features of shallow-water and supratidal evaporites: AAPG Bull., v. 53, p. 830-840.

—— 1974, Calcium sulphate minerals of evaporite deposits: their primary mineralogy, *in* A. H. Coogan, ed., Fourth symposium on salt: Northern Ohio Geol. Soc., v. 1, p. 343-348.

Krumbein, W. C., 1951, Occurrence and lithologic associations of evaporites in the United States: Jour. Sed. Petrology, v. 21, p. 63-81.

Landes, K. K., 1960, The geology of salt deposits (and) salt deposits of the United States, *in* D. W. Kaufmann, ed., Sodium chloride—the production and properties of salt and brine: New York, Reinhold Pub. Corp., p. 28-95.

Logan, B. W., et al, 1969, Late Quaternary sediments of Yucatan Shelf, Mexico, *in* B. W. Logan, et al, Carbonate sediments and reefs, Yucatan Shelf, Mexico: AAPG Mem. 11, p. 1-128.

—— R. Rezak, and R. N. Ginsburg, 1964, Classification and environmental significance of algal stromatolites: Jour. Geology, v. 72, p. 68-83.

Loreau, J. P., and B. H. Purser, 1973, Distribution and ultrastructure of Holocene ooids in the Persian Gulf, *in* B. H. Purser, ed., The Persian Gulf: Springer-Verlag, New York, p. 279-328.

Lucas, P. T., 1954, Environment of Salina salt deposition: Unpub. M.S. thesis, Univ. Michigan, 53 p.

Maiklem, W. R., D. G. Bebout, and R. P. Glaister, 1969, Classification of anhydrite—a practical approach: Bull. Canadian Petroleum Geol., v. 17, p. 194-233.

Masson, P. H., 1955, An occurrence of gypsum in southwestern Texas: Jour. Sed. Petrology, v. 25, p. 72-77.

Matthews, R. D., 1970, The distribution of Silurian potash in the Michigan basin, *in* W. A. Kneller, ed., Proc. 6th ann. forum on geol. of industrial minerals: Mich. Geol. Survey, Misc. 1, p. 20-33.

—— and G. C. Egleson, 1974, Origin and implications of a mid-basin potash facies in the Salina salt of Michigan, *in* A. H. Coogan, ed., Fourth symposium on salt: Northern Ohio Geol. Soc., v. 1, p. 15-34.

Mesolella, K. J., et al, 1974, Cyclic deposition of Silurian carbonates and evaporites in the Michigan basin: AAPG Bull., v. 58, p. 34-62.

Newell, N. D. M., E. G. Purdy, and J. Imbrie, 1960, Bahamian oolitic sand: Jour. Geology, v. 68, p. 481-497.

Nikolaev, A. V., and G. Chelishcheva, 1940, On the primary deposition of borates from sea water: Compt. Rend. (Doklady) Acad. Sci. U.S.S.R., v. 28, p. 502-504.

Nurmi, R. D., 1974, The lower Salina (Upper Silurian) stratigraphy in a desiccated, deep Michigan Basin: Ontario Petroleum Inst., 13th Ann. Conf., Paper 13, 26 p.

Ochsenius, C., 1877, Die Bildung der Steinsalzlager und ihrer Mutter-laugensalze: Halle, C. E. M. Pfeffer, 172 p.

Phleger, F. B., 1969, A modern evaporite deposit in Mexico: AAPG Bull., v. 53, p. 824-829.

Richter-Bernburg, G., 1973, Facies and paleogeography of the Messinian evaporites in Sicily, *in* C. W. Drooger, ed., Messinian events in the Mediterranean: Amsterdam, North Holland Publishing Co., p. 124-141.

Rickard, L. V., 1969, Stratigraphy of the Upper Silurian Salina Group, New York, Pennsylvania, Ohio, Ontario: New York State Museum and Sci. Serv., Map and Chart Series No. 12, 57 p.

Rusnak, G. A., 1960, Some observations of recent oolites: Jour. Sed. Petrology, v. 30, p. 471-480.

Schaller, W. T., and E. P. Henderson, 1932, Mineralogy of drill cores from the potash field of New Mexico and Texas: U.S. Geol. Survey Bull. 833, 124 p.

Schmalz, R. F., 1969, Deep-water evaporite deposition, a genetic model: AAPG Bull., v. 53, p. 798-823.

Schreiber, B. C., 1974, Upper Miocene (Messinian) evaporite deposits of the Mediterranean basin and their depositional environments: Unpub. Ph.D. thesis, Rennselaer Polytechnic Inst., 395 p.

—— and D. J. J. Kinsman, 1975, New observations on the Pleistocene evaporites of Montallegro, Sicily and a modern analog: Jour. Sed. Petrology, v. 45, p. 469-479.

Scruton, P. C., 1953, Deposition of evaporites: AAPG Bull., v. 37, p. 2498-2512.

Shaver, R. H., 1974, Structural evolution of northern Indiana during Silurian time, *in* R. V. Kesling, ed., Silurian reef-evaporite relationships: Mich. Basin Geol. Soc. Ann. Field Excursion, 1974, p. 55-77.

Shearman, D. J., 1963, Recent anhydrite, gypsum, dolomite and halite from the coastal flats of the Arabian shores of the Persian Gulf: Geol. Soc. London Proc., no. 1607, p. 63-65.

―― 1970, Recent halite rock, Baja California, Mexico: Trans. Inst. Mining Metallurgy Trans., v. 79, p. B155-B162.

―― 1971, Marine evaporites: the calcium sulfate facies: Alberta Soc. Petroleum Geologists Seminar, Univ. Calgary, 65 p.

Shlicta, P. J., 1968, Growth, deformation, and defect structure of salt crystals, in R. B. Mattox, ed., Saline deposits: Geol. Soc. America Spec. Paper 88, p. 597-617.

Sloss, L. L., 1953, The significance of evaporites: Jour. Sed. Petrology, v. 23, p. 143-161.

―― 1969, Evaporite deposition from layered solutions: AAPG Bull., v. 53, p. 776-789.

Smith, S., 1971, Discussion on paper by D. J. Shearman: Inst. Mining Metallurgy Trans., v. 80, p. B68-B69.

Stewart, F. H., 1951, The petrology of the evaporites of the Eskdale, No. 2 boring, east Yorkshire. Part II. The middle evaporite bed: Miner. Mag., v. 29, p. 445-475.

Strakhov, N. M., 1970, Principles of lithogenesis; volume 3: New York, Consultants Bureau, 577 p.

Stuart, W. D., 1971, Evaporite deposition in a layered sea—a wind-driven dynamical model: Unpub. rept., 36 p.

Treesh, M. I., and G. M. Friedman, 1974, Sabkha deposition of the Salina Group (Upper Silurian) of New York State, in A. H. Coogan, ed., Fourth symposium on salt: Northern Ohio Geol. Soc., v. 1, p. 35-46.

Tremper, L. R., 1973, Lithofacies and stratigraphic analysis of the Salina Group of the "North Slope" of the Michigan basin: Unpub. M.S. thesis, Univ. Michigan, 58 p.

Valyashko, M. G., 1972, Playa lakes—a necessary stage in the development of a salt-bearing basin (with discussion), in G. Richter-Bernburg, ed., Geology of saline deposits: Paris, UNESCO, Earth Sci. Ser., no. 7, p. 41-51.

Von Zittel, K., 1901, History of geology and palaeontology, M. M. Ogilvie-Gordon, translator: New York, Wheldon and Wesley, Ltd. and Hafner Publ. Co., 562 p.

Walther, J., 1903. Die Entstehung von Salz und Gyps durch topographische oder klimatische Ursachen: Centralbl. Mineral. Geol. Palaentol., Schweizerbart, Stuttgart, p. 211-217.

Wardlaw, N. C., 1968, Carnallite-sylvite relationships in the Middle Devonian Prairie Evaporite Formation, Saskatchewan: Geol. Soc. America Bull., v. 79, p. 1273-1294.

―― and W. M. Schwerdtner, 1966, Halite-anhydrite seasonal layers in the Middle Devonian Prairie Evaporite Formation, Saskatchewan, Canada: Geol. Soc. America Bull., v. 77, p. 331-342.

Withington, C. F., 1961, Origin of mottled structure in bedded calcium sulfate: U.S. Geol. Survey Prof. Paper 424-D, p. 342-344.

Zobell, C. E., and S. C. Rittenberg, 1948, Sulfate reducing bacteria in marine sediments: Jour. Marine Research, v. 7, p. 602-617.

Depositional Environment of Ruff Formation (Upper Silurian) in Southeastern Michigan[1]

RON BUDROS[2] and LOUIS I. BRIGGS[3]

Abstract The Ruff formation, the informal A-1 carbonate unit of the lower Salina Group, underlies the A-1 evaporite and overlies the A-2 evaporite in the subsurface of the Michigan basin. It basically is brown to dark grayish brown, fetid, unfossiliferous carbonate mudstone that ranges from limestone to dolomitic limestone and dolomite.

The Ruff formation can be divided into five lithofacies, based upon associations of subordinate lithologic constituents: microlaminated mudstone, leached mudstone, pelletal wackestone-packstone, thinly laminated mudstone, and nodular anhydrite. Deposition primarily was in shallow subtidal to infratidal, low-energy, hypersaline reducing environments, and secondarily on ephemeral tidal to supratidal flats.

In southeastern Michigan, the basal portion of the Ruff formation is intertidal-flat algal-laminated mudstone that was transgressive over the A-1 evaporite, a contiguous supratidal deposit. The remaining Ruff formation primarily is subtidal to infratidal microlaminated mudstone that grades into infratidal to intertidal pelletal wackestone and packstone peripheral to the Niagaran reefs. Deposition of these lithofacies was interrupted circumferential to the reefs by development of ephemeral tidal flats in which nodular anhydrite was deposited. The uppermost Ruff formation is the subtidal facies-equivalent of the supratidal deposit of the lowermost A-2 anhydrite.

Within the Ruff formation, differences in lithofacies distribution among various areas of the fore-slope-shelf region, and other stratigraphic variations (Mantek, 1973) are responses to different tectonic regimes within the basin. These variations have resulted in conflicting interpretations of the "reef-evaporite" relation in the Michigan basin, depending on the locations of studies.

Introduction

The Ruff formation is especially pertinent in a discussion of reef-evaporite relations, because it records hypersaline carbonate deposition within a thick evaporite sequence. Its characteristics are unique to the evaporite environment, but are typical of it. All the rocks are carbonate mudstones and algal detritus, and pellets of an unidentified crustacean are the only fossil traces. Thus, mineralogy, textures, and structures provide the keys to interpreting the environmental settings.

The Ruff formation lies within the subsurface of the Michigan basin, above the A-1 evaporite unit and below the A-2 evaporite unit (Fig. 1). These units are within the lower part of the Salina Group, which lies above and around the Niagaran pinnacle reefs (Fig. 2). The Ruff basically is a fetid, brown to dark grayish brown, stylolitic, unfossiliferous carbonate mudstone that is argillaceous or carbonaceous in places. It is limestone, dolomitic limestone, or dolomite and commonly contains thin anhydritic beds peripheral to the Niagaran pinnacle reefs. Areal extent and lithofacies distribution conform to and are controlled by topography developed during deposition of the Niagaran carbonate platform and pinnacle reefs marginal to the Michigan basin. This topography included basinal, foreslope-shelf, reef-bank, and back-reef depositional settings (Fig. 3). The Ruff formation is thickest in the foreslope-shelf region and thins toward both the basinal and back-reef regions. Tracing the formation beyond the reef bank in northern Michigan, Indiana, Ohio, and southwestern Ontario by study of wireline logs, bit cuttings, and the sparse cores is extremely difficult and uncertain; the A-1 evaporite, which provides the only reliable contrast on wireline logs, and which marks the lower boundary of the Ruff formation, is absent. Therefore, to determine lateral extent of the formation and its correlation

[1] Manuscript received March 29, 1976; accepted November 30, 1976.
[2] Continental Oil Company, Houston, Texas 77001.
[3] Subsurface Laboratory, The University of Michigan, Ann Arbor, 48109.
The writers thank Mr. Mantek and Mr. Lilienthal for help in acquiring well-log data. Special thanks are extended to Charles Kahle, Bowling Green University, for his time and constructive criticism, and to personnel of the subsurface laboratory, University of Michigan. The senior writer is grateful to Continental Oil Company for permission to publish, and to Julie LeBlanc, who typed the final manuscript.

	SERIES	GROUP	Landes (1945) & Evans (1950)	Gill (1973)	Budros (1974)
UPPER SILURIAN	CAYUGAN	SALINA	Unit H	Bass Islands Formation	Bass Islands Formation
			Unit G	G Unit	G Unit
			Unit F	F Salt	F Salt
			Unit E	E Unit	E Unit
			Unit D	D Salt	D Salt
			Unit C	C Shale	C Shale
			Unit B	B Unit / B Salt	B Unit / B Salt
			A2 Carbonate	A2 Carbonate	A2 Carbonate
			A2 Evaporite	A2 Evaporite	A2 Evaporite
			A1 Carbonate	A1 Carbonate	Ruff Fm.
			A1 Evaporite	A1 Evaporite	A1 Evaporite
MIDDLE SILURIAN	NIAGARAN	NIAGARA	Engadine	AO Carbonate	AO Carbonate
				Guelph Lockport	Guelph Lockport
			?	Clinton	Clinton

FIG. 1—Nomenclature of Salina and Niagara Groups in the subsurface of Michigan. Engadine, Guelph, Lockport and Clinton are formations. Formal status of the Ruff Formation is proposed herein.

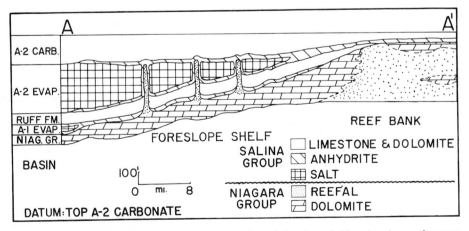

FIG. 2—North(left)-south stratigraphic cross section of foreslope-shelf region in southeastern Michigan. See Figures 3 and 4 for location of section. Modified from Gill (1973, Fig. 103).

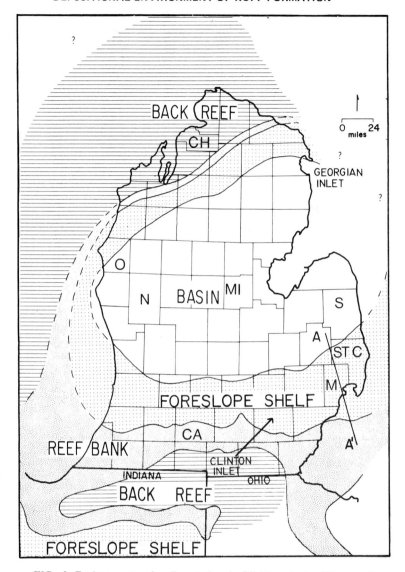

FIG. 3—Environments of sedimentation in Michigan basin, Niagaran time. After Briggs and Briggs (1974, Fig. 1). Note index of study areas: Clinton Inlet; Georgian Inlet; STC, St. Clair Co.; M, Macomb Co.; S, Sanilac Co.; MI, Midland Co.; N, Newago; O, Oceana Co.; Ca, Calhoun Co.; and Ch, Charlevoix.

near margins of the Michigan basin is tenuous and largely is a matter of stratigraphic preference (for example, see Landes, 1945; Alling and Briggs, 1961; Rickard, 1969; Shaver, 1974; Nurmi, 1974; and Janssens, 1974).

The lower contact of the Ruff with the A-1 evaporite unit in the foreslope-shelf region is considered to be unconformable, whereas the upper contact with the A-2 evaporite unit is conformable, but sharp and commonly stylolitic. In the basin both contacts are conformable.

The Ruff formation is herein proposed as the formal name of the informal unit, the A-1 carbonate. The name is derived from the type well, Consumers Power Co., 1-36

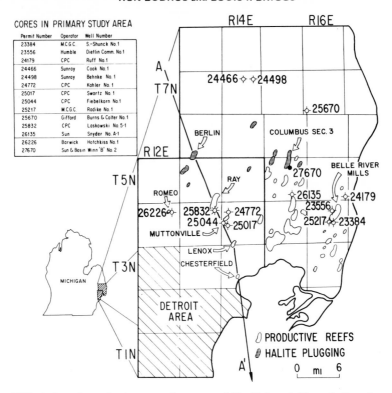

CORES IN PRIMARY STUDY AREA

Permit Number	Operator	Well Number
23384	M.C.G.C.	S-Shunck No.1
23556	Humble	Dietlin Comm. No.1
24179	CPC	Ruff No.1
24466	Sunray	Cook No.1
24498	Sunray	Behnke No.1
24772	CPC	Kohler No.1
25017	CPC	Swartz No.1
25044	CPC	Fiebelkorn No.1
25217	M.C.G.C.	Radike No.1
25670	Gifford	Burns & Colter No.1
25832	CPC	Laskowski No.5-1
26135	Sun	Snyder No. A-1
26226	Borwick	Hotchkiss No.1
27670	Sun & Bosin	Winn 'B' No. 2

FIG. 4—Location of primary study area, and St. Clair and Macomb Counties, southeastern Michigan. Base map from Mantek (1973, Fig. 3).

Ruff (SE¼, SW¼, NE¼, Sec. 36, T5N, R16E), St. Clair County, Michigan (Fig. 4). All pertinent data concerning the type well, and a complete slabbed core, are available at the Subsurface Laboratory, The University of Michigan.

Nomenclature

Landes (1945) divided the Cayugan Series in the Michigan basin into several informal units, on the basis of extensive salt beds. Units were designated by letters, from the A-unit at the bottom to the H-unit at the top (Fig. 1). In Ontario, Evans (1950, p. 59) modified Landes' classification by recognition of two salt beds in the A-unit, separated by carbonate units. In ascending order, the A-unit was divided into the A-1 evaporite, A-1 carbonate, A-2 evaporite, and A-2 carbonate. Felber (1964, p. 7) and Sharma (1966) redivided rocks of the Salina Group, based on genetic evaporite "megacycles" (a lower carbonate member and an upper evaporite member). Neither classification has gained much acceptance. Further refinements of the Landes-Evans classification were made by Ells (1967) and Gill (1973). On the basis of wireline geophysical logs, Ells (1967) divided the B-unit into a B-unit and a B-salt, and Gill (1973) recognized a basal Salina carbonate unit, the A-0 carbonate. Terminology used by Gill (1973) will be followed here, except for use of Ruff formation instead of A-1 carbonate.

Previous Studies

Most literature concerning the Ruff formation is included in regional studies of the Salina Group or in studies dealing with Niagaran reefs. Landes (1945) and Evans (1950) defined and described the Ruff formation in regional studies. Tremper (1973) included the formation in his regional stratigraphic analysis of the Salina Group in northern

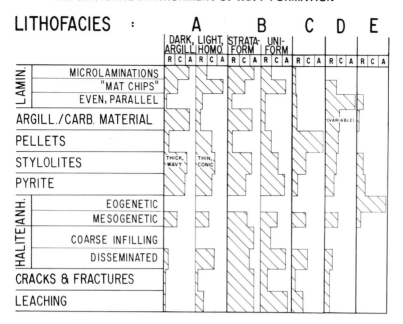

FIG. 5—Relative abundances and distributions of subordinate lithic constituents in the lithofacies A-E (R=rare, C=common, A=abundant).

Michigan, and Ells (1967) discussed relation of the Ruff to oil and gas fields of Michigan. Niagaran reefs have been the subjects of many studies, including Alguire (1962), Sharma (1966), Jodry (1969), Mantek (1973), and Mesolella et al (1974). The focus of most studies has been growth and diagenesis of reefs, and the Ruff formation is treated only as it pertains to the reef-evaporite relation.

Recently, however, as more cores have become available, some very detailed lithologic analyses of Silurian rocks of the Michigan basin include the Ruff formation. Gill (1973) investigated Niagaran-Cayugan stratigraphy, facies, evolution, and diagenesis of the Belle River Mills reef in southeastern Michigan, and Huh (1973) studied the geology and diagenesis of Niagaran pinnacle reefs in northwestern Michigan. Nurmi (1975) described the lower Salina Group, including important detailed sedimentologic and petrographic work on the salts, and Budros (1974) presented a detailed study of the stratigraphy and petrogenesis of the Ruff formation.

Location of Study Area

Primary emphasis of this investigation was on examination of cores from the foreslope-shelf region in St. Clair and Macomb Counties, southeastern Michigan (Fig. 4). However, for purposes of comparison, cores were examined that penetrated the Ruff formation in Sanilac County (basin and foreslope shelf boundary), in Midland, Newago, and Oceana Counties (basinal region), in Calhoun County (south-central foreslope-shelf region, and pinnacle reef-flank region) and in Charlevoix County (back-reef region) (Fig. 3). Locations of the cores and names of the wells are listed in Budros (1974, Figs. 9, 10, and Appendix III).

Lithofacies

The Ruff formation primarily is carbonate mudstone; it typically is dolomite or dolomitic limestone in the foreslope-shelf region, and limestone in the basinal region. However, the formation is divisible into five lithofacies based on recurrent combinations of significant constituents (Fig. 5).

Lithofacies A: Microlaminated Mudstone

Microlaminated mudstone is the most abundant lithofacies, but is highly variable in specific properties. Two end-member types are the more common. One is dark, grayish brown, argillaceous, microlaminated, characteristically dense carbonate mudstone. It may be dolomitic limestone or limestone. Faint, discrete, discontinuous, argillaceous and carbonaceous microlaminations are the most abundant constituent (Fig. 6); they commonly anastomose and coalesce into planar to wavy stylolites with thick seams of insoluble residue. The second variation is tan to brown, microlaminated dolomite mudstone, in places homogeneous, but commonly containing carbonaceous flakes termed "mat chips" (Gill, 1973) and scattered fecal pellets (Fig. 7). In contrast to the thick wavy stylolites of the dark microlaminated mudstone, conic stylolites with thin insoluble-residue seams are characteristic of the tan mudstone. Intercrystalline porosity is common, but most pore space contains disseminated halite.

The microlaminated mudstone lithofacies appears to be a subtidal-infratidal to possibly an intertidal deposit. The lithofacies type is not diagnostic of any environment, as many authors (Roehl, 1967, p. 2027; Wood and Wolfe, 1969, p. 187; Lucia, 1972, p. 177; and Gill, 1973, p. 154) have emphasized. However, in the context of the total environmental interpretation (namely, stratigraphic position, plus the presence of other, more diagnostic lithofacies in the Ruff formation) an interpretation of shallow-water deposition of the microlaminated-mudstone lithofacies in the foreslope-shelf region of southeastern Michi-

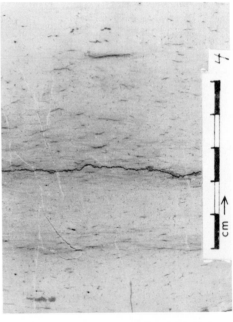

FIG. 6–Lithofacies A: Brown dolomite, microlaminated mudstone. Faint, wavy subparallel and discontinuous subparallel laminations and minute pyrite are abundant. Subhedral dolomite crystals (10-20μ) produce a predominantly xenotopic fabric. (From Consumers Power Co. 1-36 Ruff, 2569 ft [783 m])

FIG. 7–Lithofacies A: Light brown dolomite, microlaminated mudstone. Very faint, wavy, subparallel laminations and abundant discontinuous laminations ("mat chips") are present. Random replacement crystallotopic anhydrite is abundant and blocky anhydrite has filled fractures. Note flat, low-relief conic stylolite. Disseminated pyrite is common and dolomite fabric is inequigranular hypidiotopic. (From M.C.G.C. 1 Schunck, 2,512 ft [765.7 m])

gan seems to be most appropriate. More specifically, "mat chips" that were broken from more fully developed intertidal algal mats (Gill, 1973, p. 152) indicate shallow water and probable nearness to subaerially exposed flats, and argillaceous material in the mudstone suggests a quiet environment. Authigenic pyrite necessitates a reducing environment with abundant organic matter, and the presence of only algal and pelletal trace fossils indicates that the environment also was hypersaline.

Rare beds of highly distorted, light-colored mudstone resemble Holocene channel deposits on Andros Island, described by Shinn et al (1969, Fig. 6B, 6C), and channels or gullies described by Roehl (1967, p. 2032). Roehl (1967, p. 2010, 2017) stated that recognizable channel deposits in Paleozoic rocks are rare, and that in regions of Holocene arid climates, channel systems generally are poorly developed. Shinn et al (1969, p. 1226) showed that in regressive tidal-flat sequences, extensive channel systems commonly are not developed, and that transgressive depositional processes often destroy the channels.

Hence, the dark, argillaceous microlaminated mudstone is interpreted to have been deposited in a quiet, reducing, hypersaline, shallow subtidal to infratidal environment. Similar ancient deposits have been interpreted as lagoonal or quiet shallow subtidal by Thomas (1962), Textoris and Carozzi (1966), Schenck (1967), Wood and Wolfe (1969), Veizer (1970), and Wilson (1975). The light-colored mudstone, on the other hand, represents more oxygenated infratidal to possibly low intertidal deposition.

Lithofacies B: Leached Mudstone

Lithofacies B essentially is diagenetically transformed lithofacies A; consequently, environments of deposition are the same and these lithofacies commonly grade into one another. Lithofacies B is distinct and diagnostic of the evaporite carbonate mudstone. It is characterized by evidence of abundant leaching and by pore-filling halite, and in places, by abundant shrinkage cracks and fractures, in-place fragmentation and accumulations of dark, replacement crystallotopic anhydrite (Fig. 8).

The primary mechanism for generation of the diagenetic sequence possibly involves mixing of the less-saline pore waters of the Ruff formation and refluxing hypersaline brines generated by deposition of the overlying A-2 evaporite unit. Diagenetic processes were influenced to some degree by the original host rock. The dark, argillaceous and carbonaceous mudstone is transformed diagenetically into stratiform leached mudstone, whereas the homogeneous tan to brown dolomite mudstone is transformed to uniform leached mudstone. This lithofacies is discussed in detail by Budros (1974, p. 48-54, 58-59, 68-69).

The interpretation above is at variance with that of Gill (1973, p. 164), who proposed that the lithofacies represents supratidal indurated crusts, intraformational breccias, and flat-pebble conglomerates. Lithofacies B is independent of primary bedding, whereas indurated dolomite crusts would conform to or drape over the exposed beds. The carbonate textures, as seen in thin sections and scanning-electron micrographs, are also not characteristic of dolomite crusts. In addition, the leached mudstone lithofacies is distributed solely within subtidal deposits (lithofacies A) and conversely is not associated with the supratidal deposits (algal-laminated mudstone and nodular anhydrite) elsewhere in the Ruff formation.

Lithofacies C: Pelletal Wackestone-Packstone

Lithofacies C consists of various proportions of rod-shaped fecal pellets (Fig. 9) in a matrix of carbonate mudstone. Fecal pellets invariably are oriented horizontal to bedding and in some places the long axes are parallel as well, thus suggesting depositional transportation. In the foreslope-shelf region the lithofacies typically is brown dolomite and commonly is partially leached, whereas in the basinal region, it is dark limestone. This lithofacies occurs in lenticular(?) beds 1 ft (0.3 m) to 3 ft (0.9 m) thick, of pelletal wackestone and packstone that in some places appear to be cross-stratified; in other places the rocks are interbedded with carbonate mudstone (lithofacies A) and algal-laminated carbonate mudstone (lithofacies D). Recognizable burrows are not common, but structures suggestive of burrows are associated with lithofacies C. "Mat chips" and thin, discontinuous,

FIG. 8–Lithofacies B: Buff to brown, leached dolomite mudstone. Extensive shrinkage fracturing that approaches *in-situ* fragmentation in places. Fragmentation cross-cuts light, irregular, leached horizons. None of the fracturing and pseudobrecciation is apparently due to evaporite solution. All pore space (intercrystalline and interfragmentary) is filled with clear halite. (From C.P.C. 1-36 Ruff, 2,535 ft [772.7 m])

rudimentary algal laminations are within the interbedded pelletal sediments. When leached, these rocks have 10 to 15% porosity (Gill, 1973, p. 163), but commonly the porosity is local or occluded by halite or anhydrite.

Lithofacies C was deposited in infratidal and intertidal zones by soft-bodied grazers and, to a lesser degree, by burrowers. Pelleted carbonate muds are common in Holocene infratidal and intertidal sediments (Kornicker and Purdy, 1957; Ginsburg, 1957; Cloud, 1962; and Davies, 1970) and pelleted rocks in the geologic record have been interpreted to be infratidal and intertidal deposits as well (Textoris and Carozzi, 1966, p. 1384; Roehl, 1967, p. 1992; Schenck, 1967, p. 370; Kahle and Floyd, 1971, Fig. 9f; Lucia, 1972, p. 169, 172; and Wilson, 1975, SMF 16). It appears that in arid climates, pelleted deposits are common, or perhaps are more commonly preserved by hardening in the subtidal and infratidal environments; intertidal pelleted sediments are more indicative of temperate climates (Gill, 1973, p. 156). Beds or lenses of pelletal wackestone and pack-

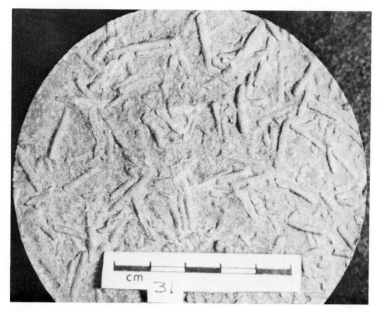

FIG. 9—Lithofacies C: Plan view of rod-shaped fecal pellets. Note near-uniformity of diameters. (From C.P.C. 1-36 Ruff, 2,612 ft [796.1 m])

stone seem to be discontinuous infratidal deposits that accumulated on minor subaqueous topographic highs. Kinsman (1966) observed that sediment on shoals in lagoons of the Trucial Coast commonly contain pellets. The pelleted wackestones and packstones interbedded with thinly algal-laminated rocks are interpreted to be low intertidal deposits. The pellets deposited in the infratidal zone may be due to benthic organisms that rework the sediment, or to reworking of pelleted carbonate muds of the tidal zones during storms (Roehl, 1967, p. 1992-2023). Davies (1970, p. 134) described pellets in carbonate-bank sediments in the intertidal levels and subtidal basal sheet, the latter apparently being derived from the levee banks during strong tides or storms.

Lithofacies D: Thinly Laminated Mudstone

Lithofacies D is characterized by abundant, closely spaced, thin, even to crinkly, parallel and subparallel, algal and argillaceous-carbonaceous laminations. Laminations in lithofacies D are thicker and are a larger proportion of the rock than in lithofacies A. Moreover, they are separated by thin beds or laminae of carbonate mudstone, which, in the foreslope-shelf region, commonly are interbedded with and grade into lithofacies A or C. Parting occurs along the laminations that have abundant argillaceous-carbonaceous material. This parting is locally called "poker-chip" parting.

Lithofacies D is made up of intertidal and subtidal deposits. Three modes of thinly laminated rocks are in the Ruff formation: (1) competent, flat to crinkly algal-laminated mudstone, (2) algal-laminated mudstone with abundant argillaceous-carbonaceous material that results in "poker chip" parting, and (3) very thinly laminated argillaceous carbonate mudstone that typically is in the basinal region and that shows pronounced "poker chip" parting.

Holocene algal-laminated sediments are produced in the intertidal and supratidal zones of Andros Island (Shinn et al, 1969), Abu Dhabi (Kendall and Skipworth, 1968), and in Shark Bay (Davies, 1970). In arid climates algal-laminated sediments generally are confined to intertidal zones, whereas in temperate climates, algal-laminated sediments are better developed in the supratidal zones. Davies (1970, p. 201) observed that flat algal-

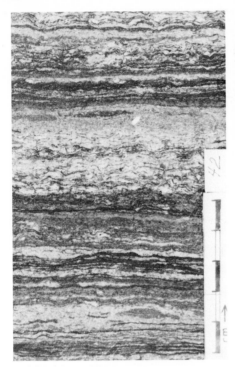

FIG. 10–Lithofacies D: Dark brown, thinly laminated dolomite mudstone. Abundant wavy to crinkly continuous and discontinuous parallel algal laminations. Probably fenestral structures are rimmed with euhedral dolomite and commonly are occluded with blocky anhydrite and hydrocarbons. (From Sun and Basin B-2 Winn, 3,121 ft [951.3 m])

FIG. 11–Lithofacies D: Brown, thinly laminated dolomite mudstone with "poker-chip" parting. Abundant, even, parallel algal and argillaceous-carbonaceous laminations. The interlaminated mudstone commonly contains abundant "mat chips," and shows an inequigranular hypidiotopic fabric. Disseminated pyrite is abundant. (From C.P.C. 1-36 Ruff, 2608.5 ft [795.1 m])

laminated mudstones with high organic content and low sediment content were high intertidal deposits, whereas flat algal-laminated mudstones with high sediment and low organic content were deposited in the low intertidal zone. Therefore, the competent, flat algal-laminated mudstones without "poker-chip" parting are interpreted as being high intertidal deposits (Fig. 10). Flat algal-laminated mudstones with "poker-chip" parting, which are interbedded with carbonate mudstone containing pellets and "mat chips," are interpreted as low intertidal deposits (Fig. 11). Numerous examples of intertidal to supratidal flat algal-laminated rocks are in the geologic record (see Budros, 1974, p. 74). The very thinly laminated argillaceous carbonate mudstones typical of the basinal region are not algal and probably represent quiet, subtidal deposits.

Lithofacies E: Nodular Anhydrite

Lithofacies E, locally called the "Rabbit Ears" anhydrite for its distinctive gamma ray-neutron response, consists of varying amounts of white to bluish white nodular anhydrite encased within brown dolomite mudstone (lithofacies A) or algal-laminated dolomite mudstone (lithofacies D) (Fig. 12). This lithofacies is highly variable, not only in the proportion of anhydrite to mudstone, but in the anhydrite fabric as well. Most variations of

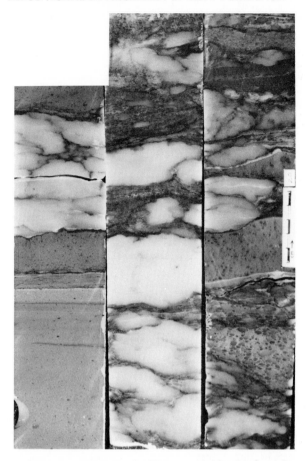

FIG. 12–Lithofacies E: White to bluish white nodular
anhydrite within brown dolomitic mudstone. Note variation
in fabric, including most types of nodular, nodular-mosaic,
and mosaic anhydrite. The larger nodules show microcrystalline and aligned-subfelted texture, whereas smaller nodules
have aligned-felted textures. (From M.C.G.C. E-1 Schunck;
left-2462 to 2463 ft.; center-2467 to 2468 ft.; right-2469 to
2470 ft.)

nodular, nodular-mosaic, and mosaic fabrics can be seen. The microcrystalline to aligned-
felted anhydrite textures are highly diagnostic of lithofacies E. Anhydrite seen elsewhere
in the Ruff formation as pore-filling and replacement anhydrite shows crystallotopic or
blocky textures.

The anhydrite is a result of eogenetic diagenesis within intertidal to supratidal flats.
The original anhydrite (gypsum?) was emplaced within unconsolidated sediments as a
result of intense evaporation of interstitial waters in the capillary zone. The occurrence of
supratidal nodular anhydrite is well documented in Holocene sediments of the Trucial
Coast, Persian Gulf, and in Laguna Madre, southwestern Texas, as well as in the ancient
record (Butler, 1969; Bush, 1973; Curtis et al, 1963; Fuller and Porter, 1969a; Ginsburg,
1975; Kerr and Thomson, 1963; Kinsman, 1966; Lucia, 1972; Masson, 1955; Murray,
1964; Shearman, 1966; West et al, 1968; Wood and Wolfe, 1969; and Wilson, 1975).

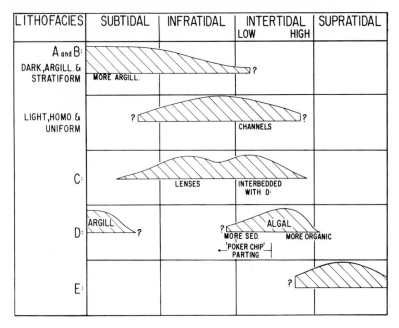

FIG. 13—Environmental relations of lithofacies A-E.

Summary of Lithofacies

Figure 13 shows a composite diagram of environmental relations of the various time-transgressive lithofacies. Whereas Gill (1973, p. 189) considers the Ruff formation as a part of cyclic sabkha sedimentation, actually, most of the formation was deposited in infratidal and subtidal environments. Supratidal deposits (lithofacies E) are a very small fraction of it. There are only a few dessication features and consequently, there are few lithoclasts and flat-pebble conglomerates, which normally are associated with sabkha lithologies. According to interpretations of lithofacies presented herein, the Ruff formation was deposited in an arid climate, primarily in subtidal and infratidal zones, in a low-energy, protected, hypersaline, reducing environment, with only limited deposition in ephemeral intertidal to supratidal flats.

In addition, normal diurnal gravitational tides may not have been the only significant factor that produced environmental realms during deposition of the Ruff formation in the shelf region; rather, tides may also have been wind-induced (Shaw, 1964, p. 63; Irwin, 1965, p. 445; Laporte, 1967, p. 84; 1969, p. 114). Tides in the shelf region could have been caused by seasonal patterns of wind direction or strength (Young et al, 1972, p. 71). Wind-driven tides occur along the Texas Gulf Coast at Laguna Madre (Masson, 1965, p. 76; Kerr and Thomson, 1963, p. 1728) and in Florida Bay (Young et al, 1972, p. 71). Hence, the term "intertidal" would not necessarily be employed in the usual sense of being between daily low and high tides, but instead could connote a "sedimentary regime that is regularly and periodically flooded by marine waters for an unspecified duration" (Laporte, 1967, p. 85). The "supratidal" zone would be topographically higher and flooded less frequently.

Distribution of Lithofacies

In the foreslope-shelf region, some lithofacies of the Ruff formation that lap upon flanks of the Niagaran pinnacle reefs contain the "Rabbit Ears" anhydrite. The general ascending sequence of lithofacies is: (1) intertidal flat algal-laminated mudstones (litho-

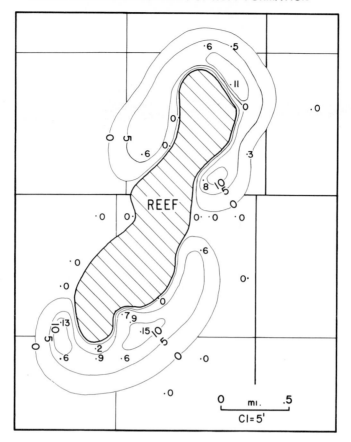

FIG. 14–Isopach map of combined "Rabbit Ears" anhydrite of
Ray field (Fig. 4).

facies D), with interlaminated mudstone and pellets dominant in some sections; (2) supratidal "Rabbit Ears" anhydrite (lithofacies E); and (3) subtidal, dark, argillaceous microlaminated mudstone, in places only partly dolomitized, that commonly grades into lighter-colored, more homogeneous dolomitic mudstone near the top of the Ruff formation. Mudstone above the anhydrite beds commonly is irregularly leached (lithofacies B) and lenses of pelletal wackestone-packstone (lithofacies C) occur sporadically. Near the reef flanks where the "Rabbit Ears" anhydrite is absent, the basal algal-laminated mudstone also is thinner, and the formation is composed primarily of microlaminated mudstone with scattered zones of leached dolomitic mudstone.

The "Rabbit Ears" anhydrite (lithofacies E) is only at the ends of the elongate pinnacle reefs and is absent along parts of the sides (Fig. 14). In addition to the Ray reef (Figs. 4, 14), anhydrite beds at Belle River Mills, Columbus Section 3, Muttonville, Berlin, Lenox, and Romeo reefs in southeastern Michigan (Fig. 4) also have similar distribution patterns (see Budros, 1974, Figs. 28-31). Distribution of the "Rabbit Ears" anhydrite adjacent to Niagaran reefs closely approximates distribution of the underlying reef talus that flanks the reefs. This can be best demonstrated at the Belle River Mills reef, where numerous cores were taken from the reef flank and off-reef lithologies.

The thicker reef talus apparently created topographically high areas that became intertidal to supratidal flats during deposition of the "Rabbit Ears" anhydrite. Stanton (1967) and Klovan (1974) stated that once a reef is established, geometry of the reef is controlled

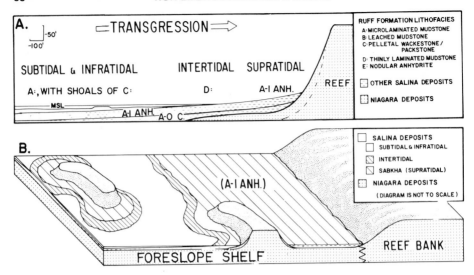

FIG. 15—Diagrammatic paleogeography of the foreslope-shelf region in southeastern Michigan during deposition of the lower portions of the Ruff formation. (A) Cross-section showing depositional environments adjacent to pinnacle reef. (B) Block diagram showing regional depositional environments in the foreslope-shelf region. (Patch reefs not included).

by the prevailing weather and the rate of subsidence. Indeed, this can be seen in modern reef complexes; on the Campeche Shelf off the Yucatan Coast there is definite elongation of reefs parallel to the dominant current. Bathymetric maps of some emergent reefs on the Campeche Shelf (Logan et al, 1969) indicate that sedimentation is dominant at the ends of the elongate reefs. Hence, once the Silurian reefs in southeastern Michigan were established, northeast-southwest elongation of the reefs and the dominant sedimentation of the Niagaran reef-talus deposits probably were controlled by: (1) currents that would appear to have been flowing parallel to the southeastern reef bank; (2) climate, including the probable prevailing southeasterly winds (Briggs, and Briggs, 1974); (3) rate of subsidence; and (4) relative compaction of reef conglomerate and tidal-flat sediments.

Vertical distribution of lithofacies in the inter-reef areas is very similar to the distribution at reef flanks where the "Rabbit Ears" anhydrite is absent, indicating a similar depositional history and a common sequence of environmental changes across the foreslope shelf.

In comparison, lithology of rocks of the Ruff formation in the basinal region is different, although the basic vertical lithofacies distribution is similar. The lower Ruff in the basinal region is fragile, very thin, argillaceous, laminated lime mudstone (locally slightly dolomitic) with pronounced "poker-chip" parting. The proportion of intercalated lime mudstone increases upward and is dominant toward the top of the formation. Most notable, however, is an interval of lime-pelletal packstone and possible algal laminations near the top of the formation; this bears a striking resemblance to the section of inter-bedded pelletal and algal-laminated rocks that is interbedded with the supratidal "Rabbit Ears" anhydrite on the flank of the Ray reef (Budros, 1974, Plates 10a and 9d), and at the top of the reef tidal-flat stage in the northern reef tract (Huh, 1973). This vertical distribution of lithofacies, including the pelletal horizon, can be seen in the Dow Chemical Salt Well 8 in Midland County, the Carter Oil 12 Lauber Well in Oceana County, and the Sun 4 Bradley well in Newago County.

Depositional History

Rocks of the lower Salina Group have been interpreted as "layer-cake" stratigraphy of alternating evaporite and carbonate deposits. But were the anhydrite and halite portions

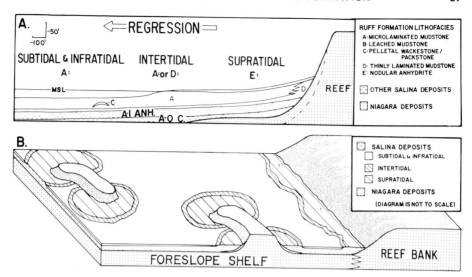

FIG. 16—Diagrammatic paleogeography of the foreslope-shelf region, southeastern Michigan, during deposition of the "Rabbit Ears" anhydrite. (A) Cross-section showing depositional environments adjacent to pinnacle reef. (B) Block diagram showing regional depositional environments in foreslope-shelf region. (Patch reefs not included.)

of the A-1 and A-2 evaporite units deposited contemporaneously, as facies equivalents, with no coeval carbonate units? Hite (1970, p. 48) stated that "all marine evaporites, unless the deposition is a result of solution and redeposition of pre-existing evaporites, have an adjacent and contemporaneous carbonate facies." Likewise, Fuller and Porter (1969b, p. 182) stated that "evaporites and carbonates deposited in two Devonian basins of western Canada occur as (1) peripheral evaporites with basin-filling carbonate or (2) peripheral carbonate with basin-filling evaporite and reefs." The first situation is also seen in Holocene deposition of nodular anhydrite in sabkhas of the Persian Gulf; this deposition is equivalent to a spectrum of seaward carbonate facies. Stratigraphic relations encourage the interpretation that evaporite beds in the lower Salina Group were deposited during a series of regressions (silling and evaporation) and the carbonates during transgressions. The units should be diachronous. Hence, it is possible that at least part of the lower Salina carbonates are contemporaneous or facies-equivalents of the lower Salina evaporites. This is especially relevant to the A-0 carbonate and the A-1 anhydrite, and to the upper Ruff formation and the lower A-2 anhydrite.

During late Niagaran time, the Michigan basin became isolated, and through evaporative drawdown described by Gill (1973), sea level within the basin dropped at least 430 ft (130 m). Thus, entire pinnacle reef masses were exposed at the beginning of the Cayugan. The A-0 carbonate and the A-1 anhydrite (possible facies-equivalents) began to be deposited and prograded basinward on what once was the Niagaran seafloor. With continued evaporation and subsidence, halite and eventually potash salts were deposited in the center of the basin, partially filling the topographic basin. Influxing normal marine waters, probably from the northeastern Georgian Inlet, the southeastern Clinton Inlet (Alling and Briggs, 1961), or from southwestern passes in the reef bank (Nurmi, 1974), transgressed and marked the beginning of deposition of the Ruff formation as thin, argillaceous, laminated carbonate mudstones in the basinal region. Intertidal flat algal-laminated mudstones and subtidal microlaminated carbonate mudstones began to be deposited transgressively over the A-1 anhydrite in the foreslope-shelf region of southeastern Michigan (Fig. 15). Regardless of the influxing normal marine waters, during deposition of the Ruff formation waters remained hypersaline, as shown by total absence of normal marine fauna. As transgression continued, intertidal flat algal-laminated mud-

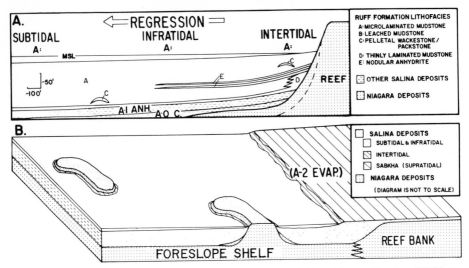

FIG. 17—Diagrammatic paleogeography of the foreslope-shelf region, southeastern Michigan, during deposition of upper portions of the Ruff formation. (A) Cross-section showing depositional environments adjacent to pinnacle reef. (B) Block diagram showing regional depositional environments in the foreslope-shelf region. (Patch reefs not included.)

stones with pellets accumulated around the exposed reefs, while elsewhere in the foreslope-shelf region, microlaminated carbonate mudstones were deposited in quiet, reducing, shallow subtidal to infratidal environments. This normal situation was interrupted by short regressive phases that produced ephemeral intertidal and supratidal flats at the ends of elongate reefs (Fig. 16). The "Rabbit Ears" anhydrite was deposited within these sediments, while microlaminated carbonate mudstone was deposited elsewhere. Within the Michigan basin, transgression of the Salina sea upon the bordering Niagaran reef bank continued after deposition of the "Rabbit Ears" anhydrite, and nearly homogeneous

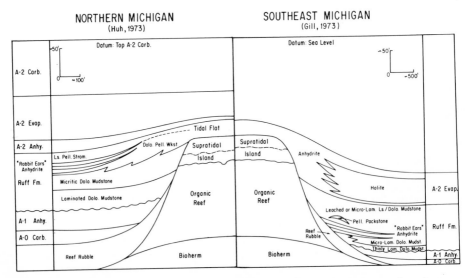

FIG. 18—Comparison of lithofacies distribution, Ruff formation, in the northern and southeastern foreslope-shelf regions. (Note different datums and scales.)

mudstone containing "mat chips" and pellets was deposited at the reef edges in shallow, oxygenated waters. Elsewhere sediments were darker, microlaminated carbonate mudstone.

Deposition of the nearly homogeneous, less argillaceous carbonate mudstone near the top of the Ruff formation in the inter-reef areas suggests that smaller amounts of argillaceous sediment were being supplied, and that the Michigan basin was again becoming isolated. Subsequently, total isolation with evaportion resulted in regression within the basin. Eventually, as sediment along the margins of the basin became exposed, the A-2 anhydrite was deposited and, with continued regression, prograded toward the basin center. The uppermost Ruff formation in the foreslope-shelf region was partly the subtidal to infratidal facies-equivalent to the prograding supratidal anhydrite (Fig. 17). Pellets that are near the top of the Ruff formation in all wells examined in the basinal region attest to shallowing of the waters throughout the basin. With accelerated evaporation and subsidence, the A-2 salt was deposited. Perhaps during this time hypersaline brines produced the leached mudstone lithofacies by brine mixing.

Stratigraphic Differences within the Shelf Region

This depositional model is not entirely applicable to foreslope-shelf regions in other areas within the Michigan basin. Elsewhere, there are differences in distribution of lithofacies within the Ruff formation and in the stratigraphic relation between the Ruff formation and the Niagaran reefs (Fig. 18). Moreover, Mantek (1973) illustrated well defined variations in width of the foreslope-shelf region, areal extent and height of pinnacle reefs, and distribution of basic lithofacies within the A-1 evaporite unit. These stratigraphic changes, including those of the Ruff formation, reflect responses to regionally different tectonics that resulted in differential subsidence in the basin (Briggs and Briggs, 1975). The tectonic and resultant stratigraphic variations have led to different interpretations of the reef-evaporite relation in the Michigan basin, depending on the location of study (see Gill, 1973, 1975; Huh, 1973; and Mesollela et al, 1975).

Summary

The Ruff formation is basically a brown to dark grayish brown, fetid, unfossiliferous calcitic to dolomitic mudstone. Five lithofacies defined by associations of lithologic constituents within the carbonate mudstone are microlaminated mudstone, leached mudstone, pelletal wackestone-packstone, thinly laminated mudstone, and nodular anhydrite. The lithofacies record deposition primarily in the shallow subtidal to infratidal zones of a quiet, protected hypersaline sea that had a reducing environment. Anhydrites represent deposition on ephemeral intertidal to supratidal flats.

In southeastern Michigan, the basal portion of the Ruff formation is intertidal flat algal-laminated mudstone that transgressed over the A-1 evaporite (the contiguous supratidal deposit). The remaining Ruff formation primarily is subtidal to infratidal microlaminated mudstone with infratidal to intertidal pelletal wackestone-packstone deposited peripheral to the Niagaran pinnacle reefs. In places, ephemeral intertidal and supratidal flats at peripheries of reefs developed, within which nodular anhydrite was deposited. The uppermost portion of the Ruff formation was deposited in subtidal environments, isochronous with supratidal deposition of the lowermost A-2 anhydrite.

Comparison of the lithofacies distribution of the Ruff formation in southeastern Michigan with that in northern Michigan (Huh, 1973) reveals significant differences. In conjunction with other stratigraphic variations noted by Mantek (1973) and Briggs and Briggs (1975), these differences illustrate regions of contrasting tectonic intensities within the foreslope-shelf region of the Michigan basin. These tectonic and consequent stratigraphic variations have led to different interpretations of the reef-evaporite relation in the Michigan basin, depending on the locations of study.

References Cited

Alguire, S. L., 1962, Some geologic and economic aspects of Niagaran reefs in eastern Michigan; *in* J. H. Fisher, chmn., Silurian rocks of the southern Lake Michigan area: Michigan Basin Geol. Soc. Ann. Field Conf., p. 30-38.

Alling, H. L., and L. I. Briggs, 1961, Stratigraphy of Upper Silurian Cayugan evaporites: AAPG Bull., v. 45, p. 515-547.

Briggs, L. I., and D. Z. Briggs, 1974, Niagara-Salina relationships in the Michigan Basin; *in* R. V. Kesling, ed., Silurian reef-evaporite relationships: Michigan Basin Geol. Soc. Ann. Field Conf., p. 1-23.

——— ——— 1975, Petroleum potential as a function of tectonic intensity in reef evaporite facies, Michigan basin (abs.): AAPG-SEPM Ann. Meeting Abstracts, Dallas, Tex., p. 7.

Budros, R., 1974, The stratigraphy and petrogenesis of the Ruff formation, Salina Group in southeast Michigan: Unpub. M.S. thesis, Univ. of Michigan, 178 p.

Butler, F. P., 1969, Modern evaporite deposition and geochemistry of coexisting brines, the sabkha, Trucial Coast, Arabian Gulf: Jour. Sed. Petrology, v. 39, p. 70-89.

Bush, P., 1973, Some aspects of the diagenetic history of the sabkha in Abu Dhabi, Persian Gulf, *in* B. H. Purser, ed., The Persian Gulf: New York, Springer-Verlag, p. 395-408.

Cloud, P. E., 1962, Environment of calcium carbonate deposition west of Andros Island, Bahamas: U.S. Geol. Survey Prof. Paper 350, 135 p.

Curtis, R., et al, 1963, Association of dolomite and anhydrite in the Recent sediments of the Persian Gulf: Nature, v. 197, p. 679-680.

Davies, G. R., 1970, Carbonate bank sedimentation, eastern Shark Bay, Western Australia, *in* B. W. Logan et al, Carbonate sedimentation and environments, Shark Bay, Western Australia: AAPG Mem. 13, p. 85-168.

Ells, G. D., 1967, Michigan's Silurian oil and gas pools: Michigan Geol. Survey Rept. Inv. 2, 49 p.

Evans, C. S., 1950, Underground hunting in the Silurian of southwestern Ontario: Geol. Assoc. Canada Proc., v. 3, p. 55-85.

Felber, B. E., 1964, Silurian reefs of southeastern Michigan: Unpub. Ph.D. thesis, Northwestern Univ., 124 p.

Fuller, J. G. C. M., and J. W. Porter, 1969a, Evaporite formations with petroleum reservoirs in Devonian and Mississippian of Alberta, Saskatchewan and North Dakota: AAPG Bull., v. 53, p. 909-926.

——— ——— 1969b, Evaporites and carbonates; two Devonian basins of western Canada: Bull. Canadian Petroleum Geol., v. 17, p. 182-193.

Gill, D., 1973, Stratigraphy, facies, evolution and diagenesis of productive Niagaran Guelph reefs and Cayugan sabkha deposits, the Belle River Mills Gas Field, Michigan basin: Unpub. Ph.D. thesis, Univ. of Michigan, 275 p.

——— 1975, Cyclic deposition of Silurian carbonates and evaporites in Michigan basin, discussion: AAPG Bull., v. 59, p. 535-538.

Ginsburg, R. N., 1957, Early diagenesis and lithification of shallow-water carbonate sediments in south Florida, *in* R. J. Le Blanc and J. G. Breeding, eds., Regional aspects of carbonate deposition: SEPM Spec. Pub. No. 5, p. 80-101.

——— 1975, Tidal deposits: Springer-Verlag, New York, 428 p.

Hite, R. J., 1970, Shelf carbonate sedimentation controlled by salinity in the Paradox Basin, southeast Utah: Northern Ohio Geol. Soc., 3d Symposium on Salt, v. 1, p. 48-66.

Huh, J., 1973, Geology and diagenesis of the Niagaran pinnacle reefs in the northern shelf of the Michigan basin: Unpub. Ph.D. thesis, Univ. of Michigan, 253 p.

Irwin, 1965, General theory of epeiric clear water sedimentation: AAPG Bull., v. 49, p. 445-459.

Janssens, A., 1974, The evidence for Lockport-Salina facies changes in the subsurface of northwest Ohio, *in* R. V. Kesling, ed., Silurian reef-evaporite relationships: Michigan Basin Geol. Soc. Ann. Field Conf., p. 79-88.

Jodry, R. L., 1969, Growth and dolomitization of Silurian reefs, St. Clair County, Michigan: AAPG Bull., v. 53, p. 957-981.

Kahle, C. F., and J. C. Floyd, 1971, Stratigraphic and environmental significance of sedimentary structures in Cayugan (Silurian) tidal flat carbonates, northwestern Ohio: Geol. Soc. America Bull., v. 82, p. 2071-2098.

Kendall, C. St. C., and Sir P. d'E. B. Skipworth, 1968, Recent algal mats of a Persian Gulf lagoon: Jour. Sed. Petrology, v. 38, p. 1040-1058.

Kerr, S. D., Jr., and Alan Thomson, 1963, Origin of nodular and bedded anhydrite in Permian shelf sediments, Texas and New Mexico: AAPG Bull., v. 47, p. 1726-1732.

Kinsman, D. J. J., 1966, Gypsum and anhydrite of Recent age, Trucial Coast, Persian Gulf, *in* J. L. Rau, ed., Second Symposium on Salt, Northern Ohio Geol. Soc., p. 302-326.

Klovan, J. E., 1974, Development of western Canadian Devonian reefs and comparison with Holocene analogues: AAPG Bull., v. 58, p. 787-799.

Kornicker, L. S., and E. G. Purdy, 1957, A Bahamian fecal-pellet sediment: Jour. Sed. Petrology, v. 27, p. 126-128.

Landes, K. K., 1945, The Salina and Bass Island rocks in the Michigan basin: U.S. Geol. Survey Prelim. Map No. 40, Oil and Gas Inv. Series.

Laporte, L. F., 1967, Carbonate deposition near mean sea level and resultant facies mosaic: Manlius Formation (Lower Devonian) of New York State: AAPG Bull., v. 51, p. 73-101.

—— 1969, Recognition of a transgressive carbonate sequence within an epeiric sea: Helderberg Group (Lower Devonian) of New York State, *in* G. M. Friedman, ed., Depositional environments in carbonate rocks: SEPM Spec. Pub. No. 14, p. 98-119.

Logan, B. W., et al, 1969, Carbonate Sediments and reefs, Yucatan shelf, Mexico: AAPG Mem. 11, p. 1-198.

Lucia, F. J., 1972, Recognition of evaporite-carbonate shoreline sedimentation, *in* J. K. Rigby, and W. K. Hamblin, eds., Recognition of ancient sedimentary environments: SEPM Spec. Pub. No. 16, p. 160-191.

Mantek, W., 1973, Niagaran pinnacle reefs in Michigan: Michigan Basin Geol. Soc. Ann. Field Conf., p. 35-46.

Masson, P. H., 1955, An occurrence of gypsum in southwest Texas: Jour. Sed. Petrology, v. 25, p. 72-79.

Mesolella, K., et al, 1972, Cyclic deposition of Silurian carbonates and evaporites in the Michigan basin: AAPG Bull., v. 58, p. 3-33.

—— et al, 1975, Cyclic deposition of Silurian carbonates and evaporites in Michigan basin: Reply: AAPG Bull., v. 59, p. 538-542.

Murray, R. C., 1964, Origin and diagenesis of gypsum and anhydrite: Jour. Sed. Petrology, v. 34, p. 512-523.

Nurmi, R. D., 1974, The lower Salina (Upper Silurian) stratigraphy in a dessicated, deep Michigan basin: Ontario Petroleum Inst. Ann. Conf., Tech. Paper No. 14.

—— 1975, Stratigraphy and sedimentology of the lower Salina Group (Upper Silurian) in the Michigan basin: Unpub. Ph.D. thesis, Rensselaer Polytechnic Inst., 261 p.

Rickard, L. V., 1969, Stratigraphy of Upper Silurian Salina Group, New York, Pennsylvania, Ohio, Ontario: New York State Mus. and Sci. Service Geol. Survey Map and Chart Ser., No. 12.

Roehl., P. O., 1967, Stony Mt. (Ordovician) and Interlake (Silurian) facies analogs of Recent low energy marine and subaerial carbonates, Bahamas: AAPG Bull., v. 51, p. 1979-2032.

Schenk, P. E., 1967, The Macumber Formation of the Maritime Provinces, Canada—Mississippian analog to Recent strand line carbonates of the Persian Gulf: Jour. Sed. Petrology, v. 37, p. 365-376.

Sharma, G. D., 1966, Geology of Peters Reef, St. Clair County, Michigan: AAPG Bull., v. 50, p. 327-350.

Shaver, R. H., 1974, Silurian reefs of northern Indiana: reef and interreef macrofaunas: AAPG Bull., v. 58, p. 934-956.

Shaw, A. B., 1964, Time in stratigraphy: McGraw-Hill Book Co., New York, 365 p.

Shearman, D. J., 1966, Origin of marine evaporites by diagenesis: Inst. Mining Metallurgy Trans., v. 75, Sec. B, Bull. 717, p. 208-215, Disc. p. B82-B86.

Shinn, E. A., R. M. Lloyd, and R. N. Ginsburg, 1969, Anatomy of a modern carbonate tidal flat, Andros Island: Jour. Sed. Petrology, v. 39, p. 1202-1228.

Stanton, R. J., Jr., 1967, Factors controlling shape and internal facies distribution of organic carbonate buildups: AAPG Bull., v. 51, p. 2462-2467.

Textoris, D. A., and A. V. Carozzi, 1966, Petrography of a Cayugan (Silurian) stromatolite mound and associated facies, Ohio: AAPG Bull., v. 50, p. 1375-1388.

Thomas, G. E., 1962, Grouping of carbonate rocks into textural and porosity units for mapping purposes, *in* Ham, W. E., ed., Classification of carbonate rocks: AAPG Mem. 1, p. 193-223.

Tremper, L. R., 1973, Lithofacies and stratigraphic analysis of the Salina Group of the "North Slope" of the Michigan basin: Unpub. M.S. thesis, Univ. of Michigan, 58 p.

Veizer, J., 1970, Zonal arrangement of the Triassic facies of the western Carpathians: a contribution to the dolomite problem: Jour. Sed. Petrology, v. 40, p. 1287-1301.

West, I. M., A. Brandona, and M. Smith, 1968, A tidal flat evaporite facies in the Visean of Ireland: Jour. Sed. Petrology, v. 38, p. 1079-1093.

Wilson, J. L., 1975, Carbonate facies in geologic history: Springer-Verlag, New York, 471 p.

Wood, G. V., and M. J. Wolfe, 1969, Sabkha cycles in Arab/Darb Formation off the Trucial Coast of Arabia: Sedimentology, v. 12, p. 165-191.

Young, L. M., L. C. Fiddler, and R. W. Jones, 1972, Carbonate facies in Ordovician of northern Arkansas: AAPG Bull., v. 56, p. 68-80.

Evaporite Cycles and Lithofacies in Lucas Formation, Detroit River Group, Devonian, Midland, Michigan[1]

R. DAVID MATTHEWS[2]

Abstract A study of individual beds of Devonian salt at Midland, Michigan, was based on descriptions of cores from closely spaced wells where few accurate subsurface data have been available. Many cycles of evaporites were defined. Cores show that salinity of depositional environments gradually vacillated from highly saline to nearly normal, resulting in gradational successions of evaporite lithofacies both vertically and horizontally. Interruptions of the sedimentary cycles, shown by missing facies, were minor, considering both map position and time.

Some of the depositional cycles have been correlated from the southeastern edge of salt beds of the Detroit River Group in east-central Michigan to the limit of preserved salt in the westernmost part of the lower peninsula of Michigan. Tentative correlations of some sulfate facies have been extended eastward to Ontario.

Conclusions are that: (1) many of the cyclic sections can be correlated; (2) correlation by cycles is a valuable adjunct to correlation strictly by lithology, and in many cases it is superior; (3) evaporite maxima and minima of correlative cyclic sections record nearly parallel time lines; (4) convergent lithologic boundaries are crossed by nearly parallel time lines in repeated instances; (5) with each cycle, space in the depositional basin was filled with evaporite lithofacies appropriate to the local salinity gradient and to map positions of gradients; (6) facies predominantly of halite changed laterally to sulfate rock with little or no halite; and (7) these strata were deposited "synchronously" and in nearly equal thicknesses across distances as great as 38 mi (61 km).

Introduction

Salt beds of the Lucas Formation, Detroit River Group (Devonian), have been mined by solution by The Dow Chemical Company at Midland, Michigan (Fig. 1), since the 1940's. Descriptions of cores and other subsurface records saved over the years were used in a study begun in 1969 of individual salt beds at the plant site. Correlation of evaporite cycles solved an immediate problem of salt-bed correlation, which proved to be related to previously unrecognized local facies change. An informal numbering system developed for cycles in the Lucas Formation at the Midland plant site has been used in The Dow Chemical Company's drilling projects as far westward as Ludington, Michigan, and eastward to near Sarnia, Ontario (Fig. 1). Numbered cycles, and the terms "upper salts," "middle salts," and "lower salts" used in this paper are strictly informal.

Basis for the Study

Study of individual salt beds was undertaken to improve the degree of confidence in correlation based on lithology alone. Descriptions of the old cores from closely spaced wells at Midland, Michigan, revealed remarkably continuous vertical successions of evaporite lithofacies in core after core, evidence that periods of increasing brine concentration were followed by periods of increasing dilution in patterns repeated many times (Fig. 2). Early in the study evidence became apparent that periods of increasing salinity or dilution could be inferred and correlated. The greatest relative salinity attained in a cycle—termed an "evaporite maximum" or "maximum"—is used herein as it was used by Andrichuk (1954); conversely the terms "evaporite minimum" and "minimum" imply the greatest relative freshening between cycles.

Correlation of depositional cycles was based on the assumption that any given cycle should have been the result of many factors acting on a single body of water. Sea-level

[1] Manuscript received March 17, 1976; accepted February 14, 1977.
[2] Minerals Department, The Dow Chemical Company, 1707 Bldg., Midland, Michigan, 48640.

The writer acknowledges the inspiration and encouragement generated by early work and personal contacts with L. I. Briggs, and the recent encouragement generously provided by L. I. Sloss. Any errors included are the sole responsibility of the writer. Appreciation is expressed to The Dow Chemical Company, U.S.A., for permission to publish this paper. A condensed version was read before the Ontario Petroleum Institute, London, Ontario, with permission of the AAPG.

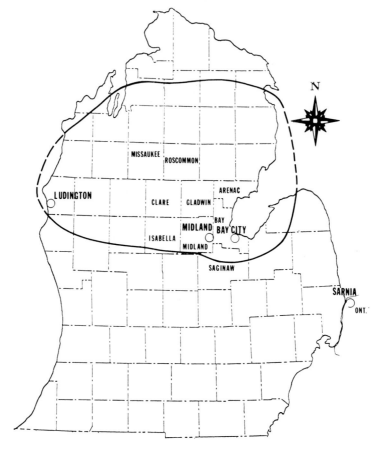

FIG. 1—Location of study area, and extent of salt in the Lucas Formation
(after Gardner, 1974).

fluctuation, degree of restriction, evaporation rate, volume of brine in the system, influx rate, rate of refluxion of brine currents out of the basin, temperature, and other factors, should have influenced cycles in a basin-wide manner. The parent brine must have vacillated gradually from highly saline to nearly normal (Fig. 3). The aggregate effect of all factors should have driven the system toward greater salinities early in the development of any individual cycle, followed by the reversed sequence during the dilution phase, late in a cycle. Ideally, this would result in definite lithologic patterns in an ascending sequence: fossiliferous limestone, nonfossiliferous limestone, dolomite, anhydrite, low-bromine halite, high-bromine halite, potash and bitterns, high-bromine halite, low-bromine halite, anhydrite, dolomite, nonfossiliferous limestone, and fossiliferous limestone. The cycle could be "reversed" at any point and the more soluble rock types would be progressively rare. The normal sequence would build to an evaporite maximum, indicated by the most soluble mineral in any given cycle. Most cycles of the Lucas Formation reached evaporite maxima in the anhydrite range, and none went beyond the stage of low-bromine halite.

At some time, the mass effect of factors influencing an evaporite cycle would change in the direction of increasing dilution. Gradual dilution would result in a complete and uninterrupted cycle, shown by a reversed vertical sequence of rock types of decreasing solubility, as described above. Andrichuk's (1954, p. 77) idealized cycle

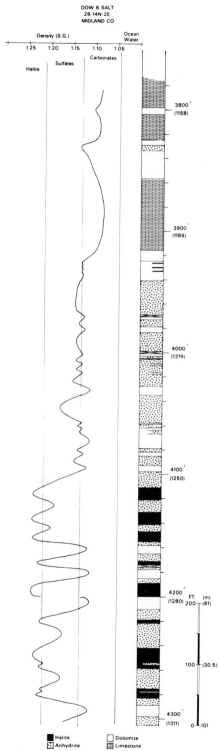

FIG. 2—Description of core, Dow 6 Salt well, showing cyclicity of evaporites.

from southern Saskatchewan shows a reversed sequence of salt, anhydrite, brown, dense and saccharoidal dolomite, and dense or fossiliferous fragmented limestone, completing the "offshore" cycle with an evaporite minimum approaching normal sea water. Gardner's (1974, p. 113) example of a cycle masimum of the Lucas Formation is overlain by a reversed sequence of anhydrite beneath dolomite.

The reversed sequences were noted by Briggs (1959, p. 52).

> The Detroit River rocks throughout the Michigan Basin exhibit many evaporite cycles that reflect progressive changes in deposition from carbonate minerals to anhydrite, to halite, and the reverse. In the rocks not containing salt, the alternation is between carbonate rock and anhydrite. Almost invariably the cycle is complete. . . .

Briggs (1959, p. 52) considered the evidence that cycles in the Lucas Formation in Michigan were "complete" and that sequences were reversed as an indication "that the basin did not completely desiccate during or following times of salt deposition, but rather the basinal waters became somewhat less saline when the balance favored greater influx over evaporation." Evaporite minerals in the upper half of the cycle are products of the freshening or dilution phase and they are precipitated from brines less concentrated than immediately earlier brines. This is true whether the section includes dolomite on anhydrite, anhydrite on halite, or halite of decreasing bromine content on potassium salts, as seen in the Silurian A-1 Salt at Midland, Michigan (Matthews and Egleson, 1974). Mechanics of origin of such evaporite deposits—the "inverse sequence" or "recessive" of Richter-Bernburg (1972, p. 35)—are not understood clearly and frequently are not discussed; yet beds of inverse sequence account for almost half the volume of evaporites in the Lucas Formation.

Although reversal of sequence is normal for evaporite cycles in the Lucas Formation, it is not the case in many other evaporites where the dilution phase left little or no record, and interrupted-cycles or "half-cycles" resulted. Flooding after desiccation would tend to build "half-cycles," as would supratidal deposition. In describing a typical sabkha cycle of beds of anhydrite over dolomite in the Rainbow Anhydrite of Alberta, Bebout and Maiklem (1973, p. 324) commented that "In some cases the sequence will not be complete vertically and may not occur laterally over the entire buildup. This lack of completeness may be a result of either nondeposition or erosion, the latter being a common feature of the subaerial supratidal environment."

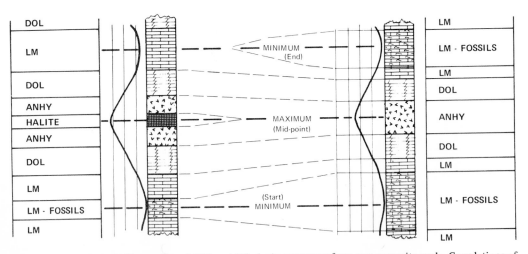

FIG. 3—Diagrammatic illustration of different lithologic sequences from one evaporite cycle. Correlations of maximum and minima define parallel time lines (broad dashed lines) that cross wedge-shaped lithologic boundaries (narrow dashed lines). At the locality on the left, greater salinity was maintained throughout.

Methods

Because The Dow Chemical Company's solution mining is located near the southern limit of the halite facies in the Lucas Formation, an effort was made to obtain records from wells more nearly central in the depositional basin, where more cycles that contain salt were expected, and where salt beds would be thicker and more likely to be noted accurately. Oil- and gas-test records in Gladwin, Arenac, Missaukee, Isabella, Bay, Midland, and Saginaw Counties (Fig. 1) were searched for wells that penetrated the salt (hundreds), that were described by detailed sample logs (a few), sonic or caliper-type logs (a few), and cores of salt above the "sour zone" (none). The Sun Oil Company 1 Mills Estate, Sec. 4, T18N, R2W, Gladwin County, about 27 mi (43 km) northwest of Midland, Michigan, was selected as the best record of a basinal well (Fig. 4A). This well became the informal "type-section" from which all evaporite cycles described in this paper were defined.

Cycles were marked on strip logs opposite midpoints in the rock section representing maximum salinity, considered to be the "maximum" of each evaporite cycle. Evidence of greatest relative freshening or decrease in relative salinity between cycles was considered to mark the position of the "minimum" ending a cycle and beginning a new one. The problem of where the minimum reversal actually occurred was ignored; for example, whether an anhydrite between two salt beds terminated an old cycle "the covering anhydrite" of Fodemski (1972), or began a new cycle, was avoided by assuming that the reversal was at the halfway point, and that deposition continued during the salinity minimum.

Correlation of evaporite cycles and lithology was started with the Sun 1 Mills, and a numbering system of the sedimentary cycles was carried throughout the cored wells. Wells with caliper logs or similar logs were included in the cycle-numbering system. From these records, and some drilling-time logs, salt beds could be located reasonably accurately, but cycles without salt could not be identified. Other types of electrical and radioactivity logs were not suited to detailed study of the evaporite cycles. Wireline logs have since been used to identify salt-bearing sections as distant as Ludington, Michigan, 120 mi (193 km) west of Midland (Fig. 4B). Salts penetrated in wells drilled by The Dow Chemical Company at Ludington are believed to be correlated with those at Midland; in any case, local correlations are excellent and although cores are not available, the concept of correlation of depositional cycles has helped in positive identification and mapping of individual salt beds in this area near the western edge of the halite facies.

In the Midland area, where core descriptions were available, correlation of cycles generally is superior to correlation based on lithology alone. In this region, where evaporite strata undergo marked lateral facies change, correlation is good, in spite of the fact that a given cycle may not be recognizable at all localities. Cycles shown by sequences of anhydrite-halite-anhydrite-halite may be "lost" if traced into a single bed of anhydrite or halite in directions normal to facies strike. Likewise cycles characterized by repetition of dolomite and anhydrite are not identifiable where shoreward facies change results in a single stratum of dolomite. Andrichuk (1954, p. 79) also cited instances in which "several discrete cycles. . .coalesce" to give the appearance of "one cycle."

Stratigraphy

Detroit River Group

In the Michigan basin, the Detroit River Group is composed of the Sylvania Sandstone, the Amherstburg Formation (the "Black lime"), the Lucas Formation, and the Anderdon Limestone, in ascending order. Several studies of the lower evaporite sections that contain "sour zone" oil reservoirs—and the productive Richfield zones somewhat lower in the section—were published by geologists of Sun Oil Company, who had developed an informal, detailed terminology (Fugate, 1968; Wirth, 1968; Sutton, 1968). Their work demonstrates clearly the numerous carbonate-sulfate-carbonate cycles below the stratigraphic position of interest for solution mining at Midland, Michigan. The general cyclic nature of evaporites of the Detroit River Group has long been recognized (Sloss, 1953; Briggs, 1959; Ehman, 1964; Gardner, 1974).

Stratigraphic terminology for the Lucas Formation was advanced by Ehman (1964). He used 119 control points in Michigan to divide the Detroit River Group above the Black lime by using mechanical log markers. He mapped the aggregate salt within four units with no attempt to identify specific salt beds other than the "Massive salt," which he called the "Big salt" or "G-H member," and which thins consistently all directions from a maximal 90 ft (27 m) in Roscommon and Clare Counties. Ehman recognized that this salt grades into anhydrite outside the central basin area, and that it is dolomitic farther from the center.

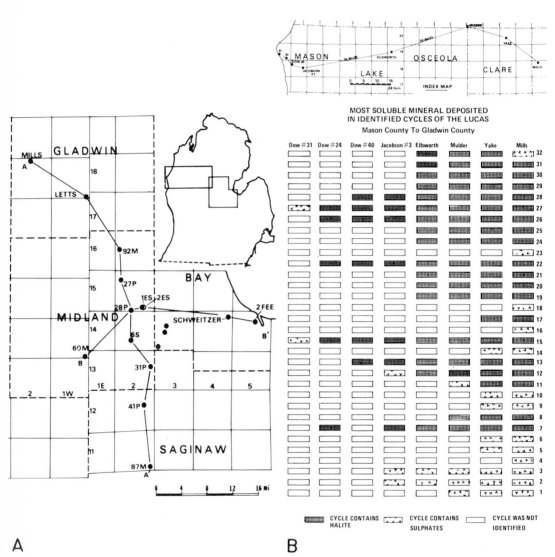

FIG. 4—A. Location map showing stratigraphic cross sections, A-A' and B-B'. The Dow Chemical Company's Midland plant is located at the site of the Dow 6 Salt well (6S). B. Chart showing the most soluble rock type in cyclic strata identified in wells located from Gladwin County to Ludington, Michigan.

In a more recent regional study, Gardner (1974) dropped the term "Anderdon" and divided the Lucas Formation into three units: the "Richfield Member" and "Freer Sandstone," the "Iutzi Member" (or "Massive anhydrite"), and the "Horner Member." The Horner member contains at least eight salt units (Gardner, 1974, p. 32); actually of the 35 or more evaporite cycles in the Sun 1 Mills well, 19 contain halite. Briggs (1959, p. 52) cited a well in Missaukee County with at least 23 evaporite cycles. In general, distances between wells in the central basin and the lack of released data on cores prevented detailed studies of the salt section above the sour zone. In this study are included data from 20 wells in Midland County (Fig. 1) in which the salt-bearing sections were cored.

Terms used by Sun Oil Company, Ehman (1964), and Gardner (1974) are shown in Figure 5, in relation to the informal cycle-numbering system used in this paper.

Richfield Zone and Massive Anhydrite

The basal unit of the Lucas Formation is the Richfield zone, which contains numerous evaporite cycles indicated by anhydrite in dolomite; this unit can be identified in these oil fields: East Norwich (Fugate, 1968), Enterprise (Wirth, 1968), Headquarters (Sutton, 1968) and Beaver Creek (Gardner, 1974). In the Sun 1 Mills well (Sec. 4, T21S, R24W), a predominantly anhydrite section at the base of the Richfield, known informally as the "Big anhydrite," was designated cycle 1 (Fig. 5). The Richfield zone is capped by the Massive anhydrite, which in the Mills well was assigned to cycles 2 and 3, although it certainly contains evidence of several depositional cycles. These cycles are discontinuous a short distance north of Midland, and can not be identified clearly at Midland.

Lower Salts or "Sour Zone"

The Sour zone "is generally considered to include those beds between the base of the massive salt and the top of the massive anhydrite." (Fugate, 1968, p. 73). At Midland, this section of rock was assigned to cycles 4 through 11 (Fig. 5). Three of these cycles (7, 8, and 11) contain salt in the central part of the basin. Salt in cycle 7 is not in any wells drilled by The Dow Chemical Company, but the cycle is indicated by anhydrite and can be traced to Dow 92 Brine well (Fig. 5). A salt bed in cycle 8 was described in the core from Dow 27 Pressure well north of Midland (Figs. 4, 5). Cycle 9 is present in the central part of the basin as anhydrite, showing salt only at Dow 92 Brine well (Fig. 5). The salt in cycle 11 is thin in the Sun 1 Mills well and does not extend into Midland County; however, an anhydrite in the stratigraphic position of cycle 11 can be traced to the Dow 31 Pressure well south of the Midland plant (Fig. 5). Other cycles below the Massive salt are stratigraphically equivalent to anhydrite in the Sun 1 Mills; these cycles may be related to salt facies northward.

Massive Salt

The Massive salt is a relatively thick, clear marker used to separate the lower salts from the middle salts. In this study, the term "Massive salt" is used, following the practice of Sun Oil Company, but Ehman (1964) and Gardner (1974) used the name "Big salt." The Massive salt is the result of four and perhaps five depositional cycles, but only two major cycles were evident at this stratigraphic position in the Sun 1 Mills well (Fig. 5). This salt may be entirely halite north of the Mills well, but in the Mills it shows an anhydritic minimum, the result of slight dilution of the parent brine. Four depositional cycles are evident within the Massive salt in the Dow 6 Salt well (Fig. 6) and at Bay City, Michigan, a fifth cycle is suspected, but is not illustrated (Fig. 7).

A few carbonate strata were of importance in this study. In most instances the rocks are dolomite, but below cycle 12 is a limestone that at Midland is evidence that halite of cycle 12 has been penetrated (Fig. 6). Beds of salt in cycles 12 and 13 are clean, but they are separated by an anhydritic minimum that is dolomitic at some localities. Salt in cycle 13 is thinner and more contaminated to the south of Midland. Salt beds of cycles 12 and 13 form a unit that is the deepest consistently mineable salt at Midland. These beds are in every well at Midland, but change southward to a pre-

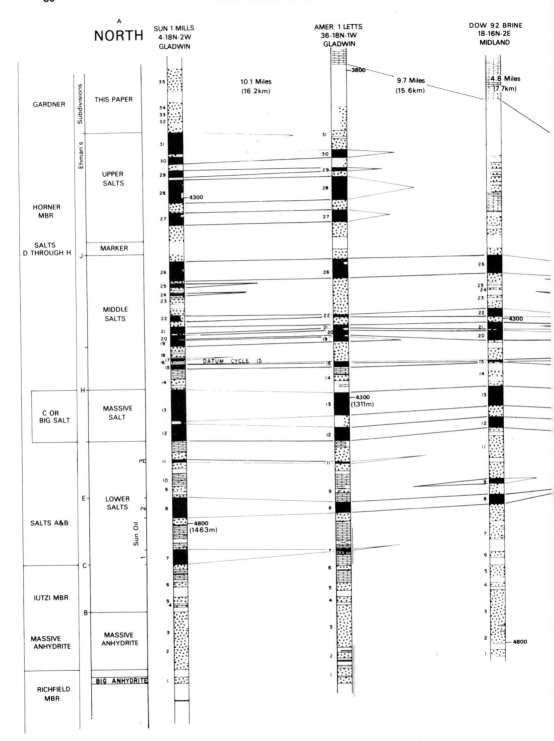

FIG. 5—Cross section A-A', north-south (Fig. 4A), showing correlation of evaporite cycles from
Gladwin County to Saginaw County.

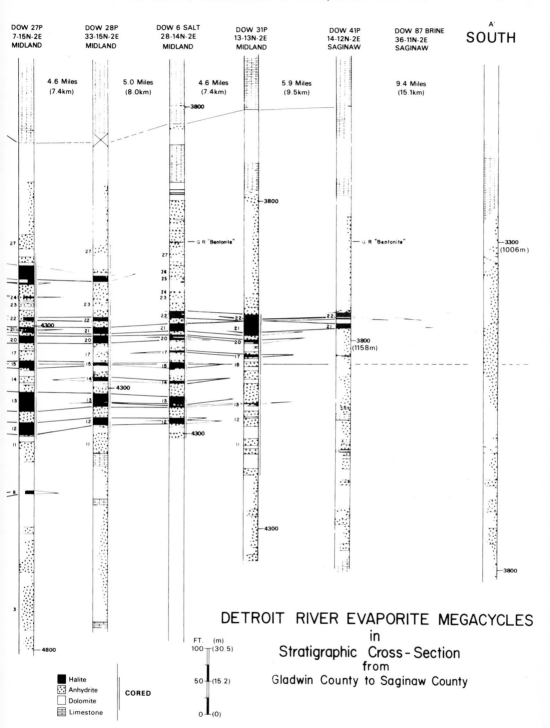

DETROIT RIVER EVAPORITE MEGACYCLES
in
Stratigraphic Cross-Section
from
Gladwin County to Saginaw County

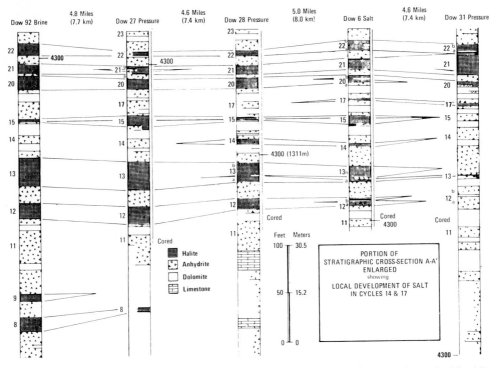

FIG. 6—Detailed portion of cross section A-A' (Fig. 4A). Massive salt contains evaporite cycles 12a, 12b, 13a, and 13b. Cycle 14 contains halite throughout only part of the total distance covered.

dominantly anhydrite facies in the Dow 31 Pressure well, where the salt of cycle 13 is only 0.1 ft (3 cm) thick.

Middle Salts

Cycle 14 is anhydrite throughout much of the area but contains halite in a few wells at Midland (Fig. 7). Salt in cycle 15 is not in all wells at Midland (Fig. 8). However, the maximum of cycle 15 is apparent in all wells, and the time line drawn through the maxima of cycle 15 is the datum for several stratigraphic cross sections (Figs. 6, 7, 9).

Cycle 16 is shown by thin anhydrite in the Sun 1 Mills, but was not identified southward (Fig. 5). Cycle 17 includes salt in the Mills well and in some wells at Midland. It is known as one of the "variable" salts (Fig. 8, 10, 11). Salt is in cycle 18 the Mills well, but it could not be traced southward. Cycle 19 is represented by salt in the Mills well; the salt can be identified only as far southward as the Amerada 1 Letts well in southern Gladwin County (Fig. 5).

A bed of salt representative of cycle 20 extends southward beyond the Dow 31 Pressure well (Fig. 5). At Midland, this salt shows evidence of being within two cycles (20a and 20b, Dow 6 Salt well, Fig. 6).

Cycle 21 contains salt as far as the Dow 41 Pressure well, and cycle 22 contains salt that terminates south of the Dow 41 Pressure. However, salt of cycle 22 shows an anhydrite minimum at Midland, and is illustration that two cycles become apparent only near the zero edge of a salt facies (Fig. 5). Salt beds of cycles 20, 21, and 22 are separated by thin anhydrite minima and the cycles thus can be traced south across Midland County. However, cycles 23, 24, and 25 cannot be identified with confidence south of the Sun 1 Mills (Fig. 5).

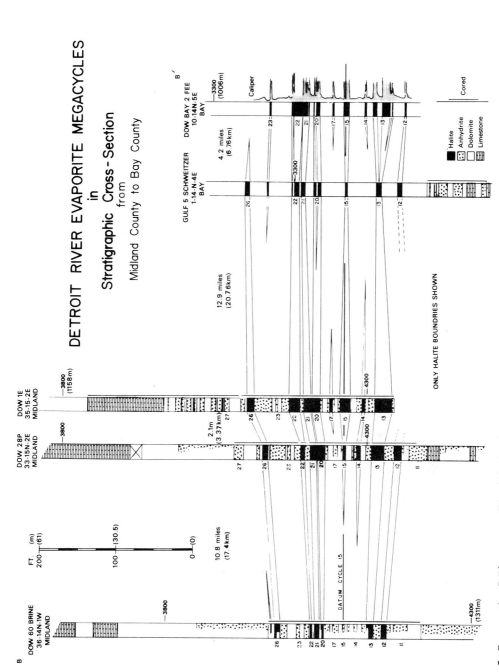

FIG. 7—Cross section B-B', east-west (Fig. 4A), showing correlations of evaporite cycles from Midland County to Bay County. Note that cycle 17 is apparent in both Dow 28P and Dow 1E, but the unit does not contain salt in the Dow 28P well. Correlation based strictly on lithology might correlate erroneously the salt in cycle 15, Dow 28P, with salt in cycle 17, Dow 1E.

A dolomite marker between cycles 26 and 27 was used to differentiate between Upper salt beds and Middle salt beds at Midland. Its position is clear only as far south as the Dow 28 Pressure well (Fig. 5).

Upper Salts

The Sun 1 Mills well shows evidence of nine cycles above the dolomite marker; five well-developed salt beds overlain by four anhydrite beds are designated as the nine maxima of cycles 27 through 35. Four of these salt beds extend as far south as the Amerada 1 Letts well, but the upper cycles are lost because of facies change and poor data (Fig. 5).

The amount of accurate (core) control is sparse in the section above cycle 22; consequently, salt beds of cycles 23, 24, and 26 were not defined as confidently as in cycles 12 through 22. The bentonite marker discovered by Baltrusaitis (Sloss, 1969; Baltrusaitis, 1974) is within the Upper salts (Fig. 5). It can be located on local gamma-ray logs, but only two wells were cored shallow enough to cut the zone, and apparently the bentonite was recovered in one core. Its position relative to the numbered cycles can be estimated only in the Dow 6 Salt well, where the core was described by C. K. Lucas (formerly of the Dow Chemical Company, written commun.) as including 6 in. of micaceous rock with very shaly partings.

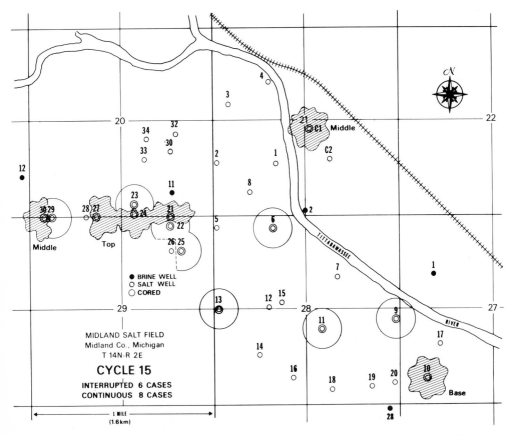

FIG. 8—Locations of wells cored through cycle 15. In wells shown by ruled patterns, salt of cycle 15 is absent.

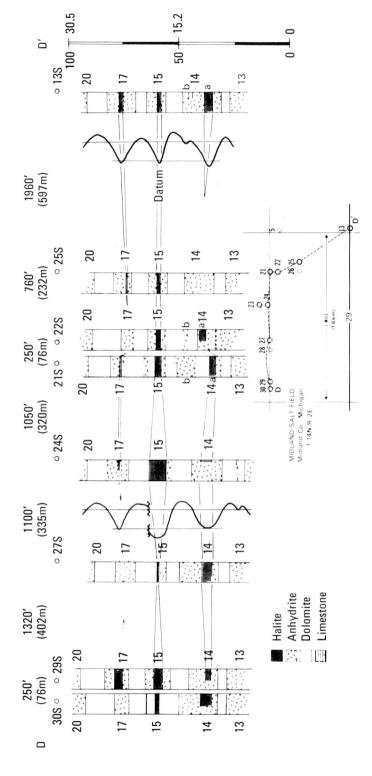

FIG. 9—Cross section D-D', east-west, showing a detailed description of closely spaced cores within part of the solution-mining area at Midland, Michigan. "Variable salts" in cycles 14 and 17 are apparent. Cycles generally are complete, as in well 13S, but a few cycles are interrupted. For example, in well 24S the salt of cycle 15 is overlain by dolomite, whereas normally it is overlain by anhydrite.

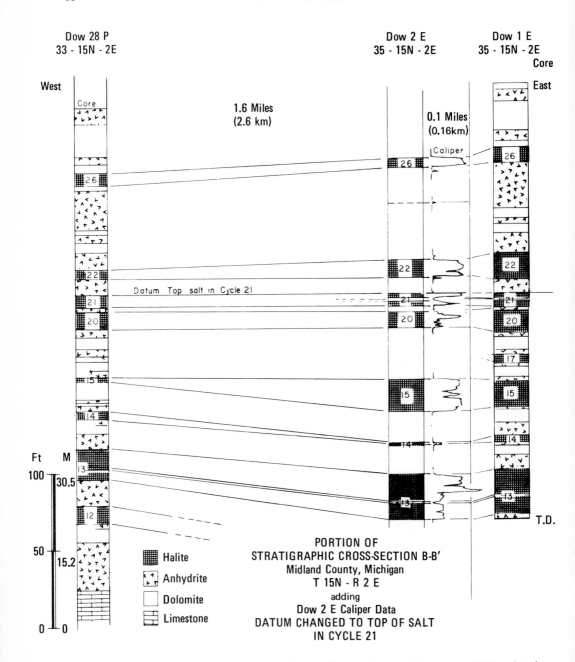

FIG. 10–Portion of cross section B-B' (Fig. 4A), with added detail of a caliper log of a nearby well. Note that the "variable" salt in cycle 17 does not appear in Dow 2E well.

Facies Changes

Continuity of Deposition

Evidence of gradual change in depositional environments is apparent in cores as an unbroken vertical succession of evaporite lithofacies; in the Dow 6 Salt well, a core 627 ft (191 m) long shows only one apparent "break" in the regular succession of rock types (Fig. 2). In seven of the cores, there are no apparent interruptions, no lithofacies are "out of place" and no clays or other clastics are distinguishable in the column. Descriptions of cores seem to document an autochthonous succession of evaporite-mineral facies, similar to the "offshore" evaporite cycle Andrichuk (1954, p. 77) described as from "an area effectively removed from the influence of source. . .of clastic material."

A few interruptions are apparent in about two-thirds of the core descriptions; these involve a missing rock type—for example, halite on dolomite without an intervening anhydrite. Some of these interruptions may be results of secondary-mineral alterations, but they are considered here to be minor, local "disconformities," representing a brief span of time when a small part of the seafloor was exposed to air or to less-dense surface waters, if the basin contained layered brines (Sloss, 1969). The small areas showing interruptions of a given cycle are a few hundreds of meters in length and width, suggesting that the conditions that led to re-solution or nondeposition were limited to small areas (Fig. 8).

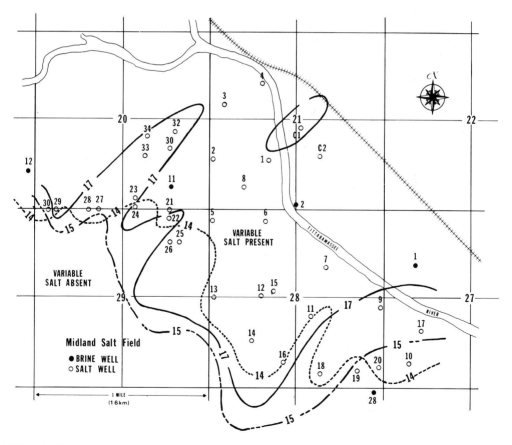

FIG. 11–Map showing edges of halite facies in cycles 14, 15, and 17, the "variable" salts, at the Midland plant site. Salt is present throughout this area in cycles 12, 13, 20, 21, and 22.

The observation that these minor "disconformities" are relatively rare and that an unbroken vertical succession of facies is normal suggests that whatever the mechanics of deposition, gradual change was prevalent in that deposition. Depth of water necessary to insure a smooth transition from one rock type to the next, whether adjacent and covering rocks were more soluble or less soluble, may have ranged from a few inches to depths approximately equal to thickness of the strata within individual cycles—no more than about 60 ft (18 m). Regardless of the manner of deposition, depositional continuity did exist in the Lucas Formation and lithologic correlations are generally straightforward from well to well, particularly along facies strike. In addition, evaporite cycles can be correlated throughout the area even in most of the cycles involving marked change in facies.

I submit that correlations of evaporite maxima define time lines, and that the correlative minima among cycles also are synchronous within about 300 sq mi (777 sq km) of Midland County, and probably across the Michigan basin. Similar conclusions were reported by Andrichuk (1954, p. 77) who stated that from southern Saskatchewan to southern Alberta, "The evaporite maxima are probably continuous."

Facies Change Within the Massive Salt

The Massive salt is an easily illustrated example of lateral extension of cycles with essentially parallel time lines that pass through nonparallel lithologic boundaries. For example, the Massive salt in the Sun 1 Mills well, Gladwin County (Fig. 4A, 5), shows evidence of two major cycles. Within 39 mi (62.8 km) from the center of the basin, the two cycles undergo the following changes (Fig. 12):

1. From midpoint to midpoint of the carbonate minima above and below the Massive salt, the two-cycle unit thins toward the margins of the basin from 92 to 79 ft (28 to 24 m) within 39 mi (62 km). This is the rate of 0.33 ft/mi (6 cm/km).

2. From the top of the basal carbonate to the base of the capping carbonate the abbreviated two-cycle unit thins from 82 to 66 ft (25 to 20 m), within 39 mi (62 km), the rate of 0.4 ft/mi (8 cm/km).

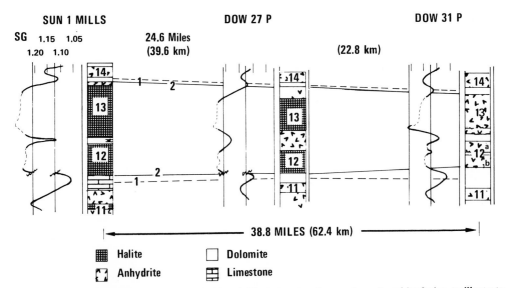

FIG. 12—Correlation of evaporite cycles 12 and 13, composing the massive salt and its facies, to illustrate the nearly foot-for-foot synchronous deposition of halite and sulfate rocks. (Locations of wells shown in Fig. 4A, cross section B-B'.) Dashed lines (1) are time lines connecting evaporite minima beginning cycle 12 and ending cycle 13. Solid line (2) connects the first and last bed of anhydrite or halite in the two cycles.

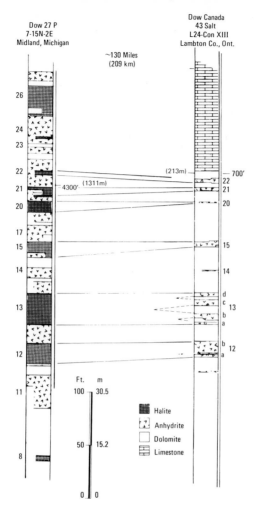

FIG. 13—Highly speculative correlations of two cores 130 mi (209 km) apart. Distance between maxima of cycles 12 and 22 is about 168 ft (51 m) at Midland; the stratigraphically equivalent(?) section at Sarnia, Ontario, is about 160 ft (49 m) thick.

3. The aggregate sulfate thickens outward from 5 to 57 ft (1.5 to 17.4 m), the rate of 1.3 ft/mi (25 cm/km).

4. The aggregate salt thins from 64 to 0.1 ft (19.5 to 0.03 m), the rate of 1.65 ft/mi (31 cm/km).

5. The two major cycles are recognized easily in cores, and are distinguishable on logs, in samples, and on drilling-time records from well to well across the 39 mi (63 km), in spite of change in facies.

Seemingly, space in the depositional basin was filled with synchronous facies appropriate to salinity gradients and map positions of those gradients. As the sulfate facies is only slightly thinner than the halite facies, foot-for-foot deposition of halite and sulfate rocks seems to have occurred "simultaneously" across a distance of 38 miles (61 km).

Cross section A-A' (Fig. 5) illustrates that the most soluble rock type in each cycle forms a lens within a larger lens of the next most soluble rock. Taken collectively, all the salt beds shown in cross section A-A' (Fig. 5) show the south half of a gross lensing pattern or wedge, as do the terminal edges of the sulfate beds, yet time lines are nearly parallel.

The outstanding fact to be learned from study of the Detroit River evaporites is that the evaporite maxima of many cycles can be correlated. This means that maximal concentration of brine in any cycle was reached at basically the same time in all places open to sea water in that cycle, and that lines constructed to pass through maxima and minima of the cycles are time lines. Evaporite rocks deposited within any cycle are of many types, which occur as sequential lithofacies forming wedges or lenses that thin out of the basin. The time lines connecting maxima and minima are very nearly parallel and these lines cross convergent lithologic boundaries as rock types change laterally.

With essentially-parallel time lines demonstrated, lateral change in rock types becomes evidence of synchronous deposition of similar thicknesses of halite, sulfate rock, and carbonate rock. Slight thinning of the section is attributed to gentle paleo-slopes and almost-uniform subsidence.

Regional Influence of Cycles

As a depositional cycle moves toward the time of maximal salinity, the salinity gradients will have moved outward from the center of the basin. Increases in salinity mark the separate transgressions of brine from basin center. Of course, gradients regressed as water freshened during the dilution phase. The depositional edge of an area influenced by the effects of a series of evaporite cycles might show evaporitic maxima by thin beds of dolomitic limestone in limestone, or by fine-grained, non-fossiliferous limestone in fossiliferous rock. The influence of evaporitic conditions seems to have gone far beyond the basinal boundaries normally shown on the maps. Gardner (1974, p. 39-40) considered the Middle Devonian anhydrite beds of northern Indiana to be laterally equivalent to the Horner member: "The maximum lateral expansion of Lucas anhydrite deposition into Indiana corresponds to. . .maximum salt extent. . ." illustrating "the migration of peripheral anhydrite facies in unison with advance environments in which halite was depositing." The evidence at Midland (cross section A-A') would suggest that the cycles likely to have had influence at a great distance from the basin depocenter would be cycles higher in the section than cycles 7 and 8, picked by Gardner. The most far-reaching cycles probably were the salt-bearing cycles 12, 13, 14, 15, 17, 20, and 21.

Correlation between the Dow 27P well at Midland, Michigan, and The Dow Chemical Company of Canada 43 Salt well near Sarnia, Ontario (Fig. 13) is based on the assumption that beds of halite and anhydrite at Midland are facies of anhydrite and dolomite at Sarnia. The Massive salt (cycles 12 and 13) comprises six cycles (at Midland there are at least four cycles) and the maxima of cycles 15, 21, and 22 are each represented by beds of anhydrite or gypsum only a few feet thick. If these speculative correlations prove to be correct, then time lines represented by evaporite maxima of cycles 12 through 22 can be shown to extend across the eastern half of the Michigan basin, bracketing lithofacies that change greatly, but the section thins only about 8 ft (2.4 m) in a total section of about 168 ft (51 m). This correlation across 130 mi (209 km) is conjectural, but near Midland, similar changes in cycles 12 and 13 have been demonstrated—with good control—across a distance of 38 mi (61 km).

Conclusions

Maxima and minima of many of the evaporite cycles in the Lucas Formation (Devonian) of Michigan can be correlated across moderately long distances. Depositional environments changed gradually in a cyclic manner that is evident both vertically and horizontally in the stratigraphic section.

The unbroken vertical succession of evaporite lithofacies is judged to be normal and interrupted cycles are considered abnormal. Maxima and minima of individual correlating cycles define time lines that are nearly parallel, and the evaporite rocks deposited between time lines connecting minima (from one set of rocks that indicate freshening of brine to the next) I regard as synchronous. Because lithofacies tend to thin from the center of the basin in long, narrow wedges, and because time lines tend to be parallel, there are numerous instances of lithologic boundaries that cross time lines. Most cycles show reversed sequences with "inverse" or "recessive" evaporites in the upper half of the cycle, so that the lithologic boundaries tend to be inclined, at both top and base.

I believe that available depositional space was filled with synchronous facies appropriate to salinity gradients in the parent brine and map positions of the gradients with time. Halite and sulfate rocks apparently were deposited simultaneously and in almost-equal thicknesses across considerable distances. Evaporites that grade outward into less-soluble facies suggest deposition in one large body of water. Absence of potassium salts and high-bromine halite requires that the mechanisms of deposition moved potassium and other bittern ions out of the basin; one such means would have been by reflux bottom currents. Reflux currents require that the brine level was only slightly less than normal sea level outside the restricted basin, to bring about a continuous flow of ions from the basin; lack of bitterns precludes massive drawdown.

The apparent filling of available space suggests that each cycle may have involved alternations of water depth ranging from a few inches, i.e., a "filled" basin, to depths equaling thickness of rocks in individual cycles. During a few cycles, water may have been as deep as 60 ft (18 m). The gradual transition of facies laterally and vertically may have been the result of advances and retreats of a sabkha-playa environment. However, I suggest that the widespread evaporite strata of the Lucas Formation were more likely subaqueous, perhaps partly deposited in very shallow water, and mostly within a large body of water.

References Cited

Andrichuk, J. M., 1954, Regional stratigraphic analysis of Devonian System in Wyoming, Montana, southern Saskatchewan and Alberta, *in* L. M. Clark, ed., Western Canada sedimentary basin: AAPG R. L. Rutherford Memorial Vol., p. 68-108.

Baltrusaitis, E. J., 1974, Middle Devonian bentonite in Michigan basin: AAPG Bull., v. 58, p. 1323-1330.

Bebout, D. G., and W. R. Maiklem, 1973, Ancient anhydrite facies and environments, Middle Devonian Elk Point basin, Alberta: Bull. Canadian Petroleum Geol., v. 21, p. 287-343.

Briggs, L. I., 1959, Physical stratigraphy of lower Middle Devonian rocks in the Michigan basin, *in* F. D. Sheldon, compiler, Geology of Mackinac Island and Lower and Middle Devonian south of the Straits of Mackinac: Michigan Basin Geol. Soc. Ann. Geol. Excursion Guidebook, p. 39-58.

Ehman, D. A., 1964, Stratigraphic analysis of the Detroit River Group in the Michigan basin: Unpub. M.S. thesis, Univ. of Michigan, 63 p.

Fugate, R. J., 1968, The East Norwich field, *in* Symposium on Michigan oil and gas fields: Michigan Basin Geol. Soc., p. 69-86.

Gardner, W. C., 1974, Middle Devonian stratigraphy and depositional environments in the Michigan basin: Michigan Basin Geol. Soc., Spec. Papers No. 1, 133 p.

Matthews, R. D., and G. C. Egleson, 1974, Origin and implications of a mid-basin potash facies in the Salina salt of Michigan, *in* A. H. Coogan, ed., General geology of salt and other evaporites: Northern Ohio Geol. Soc., p. 15-34.

—— 1975, Evaporite cycles in the Devonian of Michigan: 14th Ann. Conf., Ontario Petroleum Inst. (Chatham).

Podemski, M., 1972, Some remarks on sedimentological bases of Zechstein stratigraphy, *in* G. Richter-Bernburg, ed., Geology of saline deposits: Proc. Hanover Symp., 1968, (UNESCO), p. 219-223.

Richter-Bernburg, G., 1972, Sedimentological problems of saline deposits, *in* G. Richter-Bernburg, ed., Geology of saline deposits: Proc. Hanover Symp., 1968, (UNESCO), p. 33-40.

Sloss, L. L., 1953, Significance of evaporites: Jour. Sed. Pet., v. 23, p. 143-161.

—— 1969, Evaporite deposition from layered solutions: AAPG Bull., v. 53, p. 776-789.

Sutton, D. G., 1968, Headquarters field, *in* Symposium on Michigan oil and gas fields: Michigan Basin Geol. Soc., p. 99-114.

Wirth, J. A., 1968, Enterprise field, *in* Symposium on Michigan oil and gas fields: Michigan Basin Geol. Soc., p. 87-98.

Synchronization of Deposition: Silurian Reef-Bearing Rocks on Wabash Platform with Cyclic Evaporites of Michigan Basin[1]

JOHN B. DROSTE[2] and ROBERT H. SHAVER[3]

Abstract In the three-corner area of Indiana, Michigan, and Ohio, much of the 500-ft (150 m) Silurian section younger than the Salamonie Dolomite is a facies of the reef-bearing rocks of the Wabash platform areas of Illinois, Indiana, and Ohio. Rocks of the Salina Group and stratigraphically equivalent rocks of the platform area do not contain salts and anhydrites, but they reflect the Michigan-basin cyclic sedimentation as far south as central Indiana in the form of transgressive-regressive facies within named rock units. No major unconformity is in the section, and the up-dip carbonate rocks probably are lateral equivalents of salts in the basin.

Approximate correlation of Wabash-platform rocks in Indiana with units of the Salina in the Michigan basin is: (1) Limberlost Dolomite—lowest part of A unit; (2) Waldron Formation through Louisville Limestone—much of remainder of A unit, especially A-1 carbonate; (3) Wabash Formation—from upper part of A unit (B unit in some areas) through uppermost Salina; (4) Kokomo Limestone Member (Salina Formation)—D unit and possibly younger; and (5) Kenneth Limestone Member (Salina)—probably younger than D. Three reef-start episodes on the platform were coordinated with periods of more normal salinity during late deposition of the Salamonie, late deposition of the Louisville, early in deposition of the Mississinewa, and during deposition of the Kenneth. Some of the earliest reefs aborted during A-unit periods of above-normal salinity, including periods represented by part of the Limberlost and middle Louisville rocks, but many reefs grew during all the time of Salinan cyclic deposition.

Even in the northern platform area, where the upper part of Silurian rocks has been eroded, a complete platform buildup that included reefs could have continued to accrete during deposition of the uppermost parts of the Salina Group. These interpretations do not readily favor some current ideas on thick sabkha evaporites, hundreds of feet of drawdown, and near-desiccation in the proto-Michigan basin—nor do they favor regional development of a so-called "Niagaran-Cayugan unconformity."

Introduction

The idea that Silurian reef-building and evaporite-depositional episodes in northeastern North America were entirely sequential is entrenched in classic geological literature. Standard procedure has been to assign all normal-marine strata as high as they extend in the local section to the Niagaran Series (Middle Silurian) and to assign to the Cayugan Series (Upper Silurian) all rocks known locally to be above the designated Niagaran rocks. Thus, many rocks have been assigned to time-rock units by recognition of depositional environments. The intention was not always that, because the fossils contained were assumed to define units of time, whatever was thought about their environmental relations.

Exceptions exist. Some laminated micritic rocks that contain algal stromatolites and represent restricted environments in general were assigned to the Niagaran Series if they were known to be subjacent to normal-marine, reef-bearing rocks. Part of the Eramosa Dolomite, underlying the Guelph Dolomite of the eastern Great Lakes area, is such an example. This was correct procedure, of course, because it did not violate the type-Niagaran concept.

[1] Manuscript received November 24, 1975; accepted August 26, 1976.

[2] Department of Geology, Indiana University, Bloomington, Indiana 47401.

[3] Department of Geology, Indiana University, and Indiana Geological Survey, Bloomington, Indiana 47401.

We are grateful to many respondents who have given valuable perspective, although we are not all in full agreement. We acknowledge especially the help of A. J. Boucot, Oregon State University, concerning the stratigraphic meaning of scores of Silurian species from the lower Great Lakes area. Consultations with Richard Liebe, State University of New York at Brockport, and Lawrence Rickard, New York State Museum and Science Service, have been valuable. Much information about conodonts is from Carl Rexroad, Indiana Geological Survey, and Charles Pollock, Amoco Canada Petroleum Co., Ltd., Calgary; this work was supported partly by N.S.F. grant 5629. Publication is by permission of the Indiana State Geologist.

Another exception is a similarly arranged sequence in northern Indiana, in which some normal-marine, reeflike, stratified bodies of rock that are middle to late Cayugan are at higher elevations than nearby laminated micritic rocks which were assigned early to the Cayugan. Because of poor outcrops, the reeflike bodies were interpreted as Niagaran erosional remnants projecting through onlapping Cayugan strata. The now-identified units in this example are the Kenneth and Kokomo Limestone Members of the Salina Formation (Fig. 1); drilling has proved that the so-called "Niagaran" masses are younger than laminated micrites assigned to the Cayugan.

Evidence of the kind described above generally was used to assume that an uncon-formity separates Niagaran and Cayugan rocks virtually throughout northeastern North America (see Grabau, 1909; Cumings and Shrock, 1928; Schuchert, 1943). Examples of nomenclatural determinations shown above suggest the problem with this interpretation; alternation between normal and restricted environments, recorded by similar sequences in *both* Niagaran and Cayugan rocks, is now apparent. These prob-lems are compounded because some stratigraphers still interchange "Salina Group" (the evaporite-bearing sequence) and "Cayugan Series,"—as if penesaline to hyper-saline deposition began synchronously everywhere, and as if deposition of evaporites in the Michigan and Appalachian basins occurred only after most of the Midwest became emergent.

To summarize, the traditional rules are simple: if the strata contain abundant fossils and rocks associated with reefs, the strata are Niagaran; however, if fossils are sparse or of restricted kinds, and rocks are nonreef and partly evaporitic, the strata are Cayugan. These rules are still used by some geologists who believe that lithologic breaks must be found in local sequences, even when applying time and time-rock boundaries.

The concept of wholly sequential depositional episodes of reefs and evaporites was challenged in varying degrees of directness by Sloss (1947), Lowenstam (1950), Liberty and Bolton (1956), Alling and Briggs (1961), and others. Much evidence now shows the correctness of the challenge. It includes information from many hundreds of subsurface data points in the Michigan, Illinois and Appalachian basins, and in what once was a large Silurian shelf area, the Wabash platform. The platform was larger than crestal areas of present arches and it separated the Silurian prototypes of the basins. New evidence comes also from integration of older data from outcrops with data from the subsurface, and from integration of classic and newer paleontologic information, particularly concerning zonal brachiopods and conodonts. Detailed facies relations now can be recognized from basin to platform and from reefs to interreef areas, in-cluding areas of deposition of evaporites.

However, part of the new evidence and opinions—particularly from studies of pinnacle reefs and evaporites in the Michigan basin—appears to contradict the asser-tions made above and to revive the ideas of wholly sequential episodes of reefs and evaporites, and of a widespread Niagaran-Cayugan unconformity. The contradictions are based mainly on regional projections from geochemical and sedimentologic obser-vations within limited parts of basins. We believe that relatively fine-scale interpreta-tions of a sedimentational sequence around a given basin reef, or group of basin reefs, and for limited stratigraphic intervals, have been applied incorrectly to unrelated sedi-mentational events at the very large scales of Midwestern geography and of the total reef-stratigraphic range.

Scope and Purpose

We seek reconciliation among interpretations of events on the platform and in the basin by integrating paleontologic and environmental data with rock-body geometry on a regional scale. We offer four main propositions:

(1) No Niagaran-Cayugan unconformity worthy of so impressive a title exists in the Great Lakes area.

(2) The oldest Salina sediments in the Michigan and Appalachian basins, including some evaporites, actually are Niagaran in age, because they predate much of the out-cropping Guelph Dolomite of western Ontario and the Oak Orchard Dolomite (upper type Niagaran) of western New York (Fig. 1).

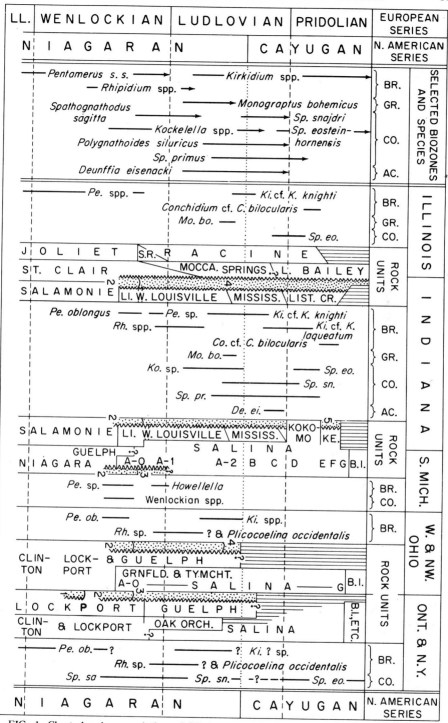

FIG. 1–Chart showing correlation of Silurian rocks of Great Lakes area, ranges of key fossil species and zones, principal reef starts (stippled; numbered 1-5), and abortions (closed tops, stippled areas). Symbols: BR.-brachiopods; GR.-graptolites; CO.-conodonts; AC.-architarchs; B.I.-Bass Islands; KE.-Kenneth; LI.-Limberlost; LL.-Llandoverian; S.R.-Sugar Run; W.-Waldron.

(3) All the Silurian reefs of the Great Lakes area did not begin and abort at the same time; more than five Middle and Upper Silurian reef generations exist altogether, some of which were coordinated in growth and abortion with carbonate-evaporite cycles in the evaporite basins. Some reefs were very short lived, including pinnacle reefs of the Michigan basin; others persisted probably until Early Devonian time, and therefore existed as growing structures during deposition of Silurian evaporites—although not necessarily in close proximity.

(4) Continued misuse of time and time-rock nomenclature has resulted in serious misinterpretation of tectonic-sedimentational events on an interregional scale and of reef-evaporite relations in general. The geographically designated series terms (e.g., Niagaran) should be used in the time-correlative sense as close as possible to the proposal of Rickard (1975), until a more refined proposal is made. The spatially designated series terms preferred by many persons (e.g., Middle Silurian) are used here in accord, and we also use the European series terms (e.g., Wenlockian). The European terms have become common for rocks that have index fossils pertinent to classification.

Structure, Thickness, and Outcrop Relations of Reefs

The top of mostly nonreef rocks of the Lockport Group of northwestern Ohio and of equivalent rocks in Indiana (Fig. 2) is a structural datum to which nearly all the reef-stratigraphic intervals discussed here can be keyed. In Indiana the contoured horizon is the top of the Salamonie Dolomite (Fig. 1). The datum also is the bottom of brown carbonate rocks that constitute the Limberlost Dolomite in central northern Indiana. In northeastern Illinois this horizon is very nearly or exactly the top of the Joliet Dolomite; in the deeper Illinois basin, in parts of Illinois and Indiana, it is the

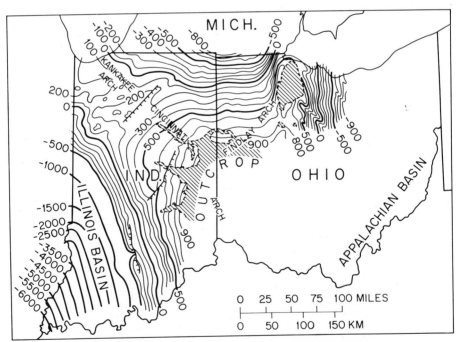

FIG. 2–Structural contour map. Contour interval 100 ft (30.5 m). Datum is top of pre-reef part of Lockport Group in northwestern Ohio, and of Salamonie Dolomite in Indiana. Salina Group lies next above or closely above datum in Michigan basin. Reef-bearing rocks lie above datum in most of Indiana. The part of map in Ohio provided by Arie Janssens, Ohio Geological Survey.

top of the St. Clair Limestone (Fig. 1). In southern Michigan, the datum is nearly or exactly the top of the Niagara Group (top of undifferentiated Niagaran, or so-called Gray and White Niagaran).

Datum points for this horizon are based mostly on cuttings and cores. The whitish (pure) carbonate rocks below this horizon are coextensive with similar rocks that make up a near-top part of the massive reef or bank in southern Michigan (Fisher, 1973). The phrase "nearly or exactly" is used above in reference to extension of the structural datum beyond the area shown in Figure 2, because a few tens of feet of brown carbonate rock above the datum in Michigan have been classified as part of the Niagara Group or as part of the Salina Group. Moreover, exact correlation of the thin Sugar Run Formation (above the Joliet Formation, Fig. 1) in northeastern Illinois is uncertain.

The structural horizon does not represent an unconformity in the area shown by Figure 2, but it probably is modestly time-transgressive along the basin margins. For example, in the Illinois basin of southwestern Indiana, the top of the St. Clair rises through the interval represented toward the platform by the Louisville Limestone; this means that one or both rock units are time-transgressive in this area. In contouring, discrete reefal buildups were ignored; many extend from and far above this structural horizon.

These buildups compose one of the largest generations of reefs in the Great Lakes area (second generation, Fig. 1). Thus, roots of the multistage pinnacle reefs of southern Michigan are at and below the critical structural datum, as are roots of many reefs in the outcrop area of the lower Great Lakes. The small Gasport reefs of New York (Crowley, 1973) and adjacent Ontario (the oldest reefs described by Pounder; 1963a, b) appear to represent an earlier generation (first generation, Fig. 1). Older buildups in the Mayfield and Byron Dolomites of Wisconsin, in the Hopkinton Dolomite of Iowa, and in rocks of similar ages on Manitoulin Island, Ontario, are not considered in the scheme of Figure 1.

The contoured top of the Salamonie (Figs. 1, 2) is exposed at elevations near 900 ft (270 m) along the Ohio-Indiana line and descends to about 200 ft (60 m) along the Kankakee arch in northwestern Indiana. There some large exposed reefs have been drilled to depths of approximately 400 ft (120 m) without interruption in the reef structure. Their bottoms are below the top of the Salamonie. These reefs may have been top-eroded a few hundred feet, but nevertheless they are almost equal in size to the buried pinnacle reefs of Michigan. Larger reefs are present southward from northwestern Indiana, where they generally have been eroded less (note thickness of reef-bearing strata shown in Figure 3). Assuming that the proto-Michigan basin subsided more rapidly during growth of pinnacle reefs than did the Wabash platform to the south, the overall platform-reef thickness should represent significantly more time than does the similar thickness of pinnacle reefs in Michigan.

Understanding the relation among the small exposed reefs near the Ohio-Indiana line (from Linn Grove to Buckland, Fig. 4) and the larger reefs mentioned above in the Michigan basin and northwestern Indiana (northwest of Delphi, Fig. 4) is dependent upon this structure-thickness analysis. They belong to the second generation (Fig. 1). Pinnacle reefs of the southern Michigan basin could have begun to grow somewhat earlier than the others, however. They and many of the reefs near the Ohio-Indiana line were aborted relatively soon after inception.

The small size of reefs of western Ohio and eastern Indiana, in comparison with those of the southern Michigan basin, is due mostly to slower subsidence of the area, not to greatly different times of abortion. The smaller size of these eastern reefs in comparison with those of northwestern Indiana is due to platform-reef geometry (laterally expansive with upward growth), along with erosive and abortive truncation of the eastern reefs at a relatively low stratigraphic level (at or near the Lockport-Salamonie top, Fig. 4).

Thus, the bedrock surface differentially truncates the reef-bearing strata in the lower Great Lakes area, so that reefs of the large second generation are exposed for examination at their bottom-most and topmost stratigraphic levels from western Ohio

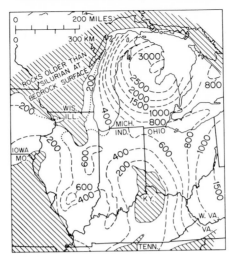

FIG. 3—Thickness of Middle and Upper Silurian rocks (including reef- and evaporite-bearing units). Contour interval 200 ft (61 m) where thickness is less than 1,000 ft (305 m), 500 ft (152 m) where thickness is more than 1,000 ft (305 m). Contour line dashed below Devonian cover, dotted where there is no Devonian cover.

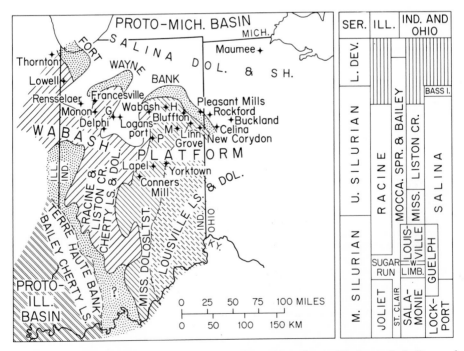

FIG. 4—Pre-Middle Devonian geologic map and locations of some Silurian reefs, Indiana and nearby areas. Symbols: G - Georgetown; H - Huntington; P - Pipe Creek, Jr.; M - Montpelier; W - Waldron; Limb. - Limberlost.

to northwestern Indiana. This geographic-stratigraphic-structural relation is shown by Figures 2 and 3, and by the pre-Middle Devonian geologic map of Figure 4. (Effects of post-Devonian erosion are not shown in Figure 4.) Reefs of other generations can now be placed in perspective with this discussion; evidence of other kinds is needed to support the stated age relations among these generations, and between reefs and evaporite-bearing rocks that are closely associated with one another.

Paleontology and Age of the Platform Reef Section

The pentameracean brachiopod taxa shown in Figure 1 have been observed repeatedly in Indiana in the vertical order and ranges shown, but they have not been recorded as completely in Illinois, Ohio, and Ontario. These brachiopods were responsive environmentally as reef-flank and near-reef faunas, but the separate ranges depicted also resulted from organic evolution; therefore, the ranges have time value. The accurate coordination (10 to 30-ft [3 to 9-m] limits) of positions of these indices within a 500-ft (150 m) section of rock in Indiana—the internal sequential relations of which are based thoroughly on physical stratigraphic methods—thus obviates the shortcoming of many so-called Niagaran indices.

The lower part of the brachiopod zonation of concern begins with *Pentamerus* (mostly *oblongus*) in rocks of the uppermost Salamonie to lowermost Louisville (Fig. 1). This is a higher interval than that part of the overall range of *Pentamerus* that characterizes the Kankakee Dolomite of Illinois and the Fossil Hill Formation of Ontario. The genus is not known to be younger than Wenlockian, and nearly all its distribution in the Great Lakes region appears to be no younger than about middle Wenlockian (middle Niagaran).

Pentamerus is succeeded, probably with small overlap, by *Rhipidium*, which is found in the Louisville Limestone. *Rhipidium* is succeeded by two species of *Kirkidium*, respectively in the upper Louisville to lower Wabash Formation, and in the upper Wabash, and by species of *Conchidium* in the upper Wabash and the Racine Dolomite, and in the Kenneth Limestone Member of the Salina Formation.

The fact that four or five such zones are vertically disposed suggests the elapse of a long period of time. This fact should be considered together with evidence drawn from conodonts. However, the standard European conodont zones have not been fully established in the Great Lakes area. The highest European zone, the *Spathognathodus eosteinhornensis* Zone, has been considered to represent latest Ludlovian (early Cayugan) and Pridolian (Cayugan) time, but the name-giver may have somewhat greater range in North America. *Spathognathodus eosteinhornensis* is reported from an unnamed shale that has been assigned to the upper Moccasin Springs Formation or the lower Bailey Limestone in the Illinois basin (Fig. 1) (Collinson and others, 1967; Berry and Boucot, 1970). It is also reported from the Kokomo Limestone Member, Salina Formation, in Indiana (Shaver et al, 1971) and from the D salt, Syracuse Formation, in New York (Berry and Boucot, 1970; Rickard, 1975).

The *Polygnathoides siluricus* zone, below the *S. eosteinhornensis* Zone, has been identified in the lower 30 ft (9 m) of the Liston Creek Limestone Member (upper part, Wabash Formation) at the type section (Pollock and Rexroad, 1973). This zone indicates a middle to late Ludlovian (late Niagaran to early Cayugan) age. As much as 250 ft (76 m) of reef-bearing rocks in Indiana are stratigraphically above this position. Other zonal conodonts are recorded in both reef-bearing and Salinan rocks in northern Indiana (Fig. 1; and Shaver et al, 1971).

Two significant graptolite and acritarch zones are those of *Monograptus bohemicus* and *Deunffia eisenacki* (Fig. 1). *M. bohemicus* (middle Ludlovian; late Niagaran) is in the Mississinewa Shale Member (lower Wabash Formation) of Indiana (Cumings and Shrock, 1928) and is near the top of the Moccasin Springs in a part of Illinois where the top of the Moccasin Springs also is the Silurian-Devonian unconformity (Ross, 1962). *Deunffia eisenacki* (Ludlovian; late Niagaran - early Cayugan) is in the middle to upper Mississinewa Member of Indiana (Wood, 1975).

All these observations support an age of normal-marine reef-bearing rocks in the southern Great Lakes area that ranges from middle Wenlockian (middle Niagaran) well into the Pridolian (Cayugan). The Pridolian age could apply even to eroded upper parts of the largest reefs in the Indiana-Illinois area southward from Lake Michigan; it surely applies to upper parts of the slightly eroded or uneroded reefs farther south, some of which probably represent latest Silurian time (Fig. 1). Further, these observations—applied to outcropping reef-bearing rocks in Ohio and westward to north-central Indiana—show that the traditionally assumed Niagaran-Cayugan unconformity ranges from middle Wenlockian well into the Pridolian. No such single unconformity exists, whatever separate, minor unconformities may apply around some periodically emergent reef tops. (See facies relation proposed by Janssens [1974] and Shaver [1974b], for rocks embracing many different stratigraphic levels of what has been considered as a single unconformity.)

Correlation of Midwestern reef-bearing rocks with the section in western New York remains somewhat problematic. The standard Niagaran-Cayugan boundary is to be "taken as the top of the Niagaran (top of Lockport Group) in the Niagara Gorge" (Rickard, 1975, p. 3); that is, *neither* as the top of the Guelph elsewhere, *nor* as the base of some lower Salina rocks distant from the gorge that are late Niagaran in age. Zenger's (1965) assessment of type-Lockport faunas, considered with probable *Rhipidium* in the Guelph of Ontario and doubtful *Kirkidium* in the Oak Orchard Dolomite (highest unit of the Lockport) of western New York, suggests an age of about middle Ludlovian for youngest type-Niagaran rocks. Partly unpublished information on conodonts from the uppermost Oak Orchard at Rochester, New York, 75 mi (120 km) from the gorge, (Rickard, 1975, and oral communication, 1975; Richard Liebe, written communication, 1975) suggests that the Niagaran-Cayugan boundary might be in upper Ludlovian rocks.

Nevertheless, faunal associations and the Lockport nomenclatural scheme used on a physical basis as far west as the eastern border of Indiana show that parts of the Lockport older than the Oak Orchard and Guelph in New York and Ontario correlate with and below the lowest part of the reef-bearing section of the Wabash platform. Also, the type Oak Orchard and Guelph in Ontario are partly, if not mostly, the same age as the Louisville Limestone, which is in the lower part of the reef section of Indiana. Finally, an upper part of the reef section in Indiana, Illinois, and (?)Ohio is Cayugan in age (Fig. 1).

Ages of Reefs and Upper Niagaran Strata, Michigan Basin

The youngest reliable guide fossils in pinnacle-reef strata in Michigan are *Pentamerus* s.s. (no younger than Wenlockian; middle Niagaran) and some Wenlockian conodonts (Mesolella et al, 1974, and Mesolella, oral communication, 1975). *Pentamerus* also was reported by Sharma (1966). Felber's (1964) report of *Conchidium* appears to contradict the Wenlockian age, but in the taxonomic terms employed here, this fossil hardly could be *Conchidium*; rather, it could be *Rhipidium*, or another pre-*Kirkidium* pentameracean (which could be *Pentamerus* itself), or earliest *Kirkidium*.

Although Mesolella et al (1974) suggested (on the basis of a *Howellella* brachiopod fauna) that the upper, algal-stromatolitic parts of the basinal pinnacle reefs could be as young as Pridolian (Cayugan), this age is most improbable. The cited fauna, known widely and including *H. corallinensis*, ranges from middle Wenlockian into Pridolian (middle Niagaran into Cayugan) (Cumings and Shrock, 1928; Williams, 1919; Berry and Boucot, 1970; Boucot, written communication, 1961, 1963, 1965). Thus, *Howellella*, as cited, represents an environmental circumstance without a refined age indication.

The faunal evidence from the basin, taken with the fact of overtopping of pinnacle reefs by the A-2 Salina carbonate (at least partly Ludlovian, Fig. 1), indicates only a possibility that any part of the pinnacles is as young as early Ludlovian (late Niagaran). If Gill's (1975) reef-stratigraphic proposals are correct, the pinnacle reefs entirely predate the A-0 carbonate and could be no younger than Wenlockian; if the views of Mesolella et al (1975) are correct, the reefs have an A-1 carbonate component, and their upper parts could be as young as early Ludlovian.

The terms "Guelph Dolomite" and "Niagara Group" are used variably in the interior basin, often for rocks locally below strata designated in rock-unit procedure to be basal Salina. This includes the reefs proper, sometimes the carbonate rocks called "Brown Niagaran" (part(?) of A-0 unit of Gill, 1975), and sometimes even the algal-stromato-litic tops of reefs. The age of these Guelph (Niagara) rocks with respect to the out-cropping Guelph of western Ontario has not been clear. Nevertheless, almost invariably the designators of Guelph tops at *different* stratigraphic positions—both reef-related and off-reef—seem to believe that they also are designating a Niagaran-Cayugan bound-ary, which is then used to coordinate paleogeographic and tectonic-sedimentational reconstructions.

Nearly all these designations are below the actual type-series boundary, but some are near the boundary along the southern flank of the basin. Where the lower Salina salts and anhydrites reach zero thickness toward the platform, stratigraphers tend to assign the remaining A-carbonate rocks to the Niagara Group. This has been true especially where geophysical logging was used. It was pointed out by Mesolella et al (1974) and by Mesolella (presentation in symposium for this volume), and is borne out by our studies and those of Fincham (1975) and Fincham and Fisher (1975). Fincham and Fisher (1975) proposed that the eastern part of the Fort Wayne bank in Indiana could range from Salina A to Salina C in terms of relative age. We agree. In the wells of Indiana and Michigan that are common to our studies, their designated Niagaran tops rise from just above the top of the Salamonie Dolomite to a position within the Missis-sinewa stratigraphic interval (Fig. 1). In places, for example, a part of our lower Salina is their A-1 carbonate unit as expectable, but in other places, the same strata are within their upper Niagara. Their A-2 carbonate, also a part of our Salina, ranges into the Mississinewa position (see also Okla, 1976).

This commentary notes a real achievement toward understanding basin-platform relations and the remarkable facies between evaporite- and reef-bearing rocks. The obstacle to such understanding really has not been selective designation of what shall constitute given rock units; it has been and continues to be application of the dictum "Salina = Cayugan" and "Niagara = Niagaran," which simply defines the facies relation out of existence.

Physical Stratigraphy and Sedimentation

In the Wabash platform area the earliest beginnings of reefs are in upper parts of the St. Clair Limestone, Salamonie Dolomite, Niagara Group, and the part of the Lockport Group older than the Guelph (Figs. 1, 5); also, they are in the lower part of the sequence made up by the Sugar Run and Racine formations. The southern platform margin (bordering the proto-Illinois basin) evidently was stable, as judged from a bar-rierlike feature, well over 500 ft (150 m) thick in places, named the Terre Haute bank. The proto-Illinois basin was somewhat starved, and no or few evaporites were deposi-ted, which accounts for the thin off-reef section (Figs. 3, 5, 6) that contrasts with the thick off-reef evaporite-bearing section in the proto-Michigan basin. Lower Devonian rocks, immediately above the off-reef section in southwestern Indiana and adjacent Illinois, make up the upper part of the Bailey Limestone, which has no known internal unconformity.

The proto-Michigan basin had an unstable southern margin that was marked by two successive bank systems, the older in southern Michigan (Fisher, 1973), and the mostly younger Fort Wayne bank in northern Indiana (Fig. 7). From the alternating carbonate and salt-anhydrite basinal sequence, salt and anhydrite are lost toward the platform, but the carbonate rocks, including salt equivalents, extend onto the Wabash platform and there are part of the overall reef-bearing section.

The Fort Wayne bank was open or nonexistent in places, particularly during its early history. Thus, up-dip Salinan and Salina-equivalent carbonates, which are light to dark-colored, fine-grained to micritic dolomites—even thinly laminated shaly appearing dolomites—extend well into the central platform area to central Indiana. The lower-most conspicuous example includes parts of the Limberlost Dolomite, the Waldron

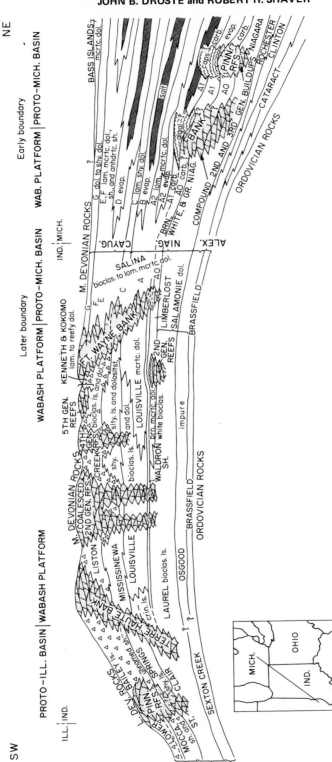

FIG. 5—Diagrammatic cross section showing restored Silurian section from proto-Illinois basin, across Wabash platform, into proto-Michigan basin. Exact age relations of pinnacle reefs and lower Salina units in Michigan are uncertain.

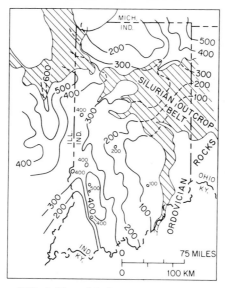

FIG. 6—Map of Indiana and bordering areas showing thickness of Salina Formation (upper Niagaran through Cayugan) in northern Indiana and Salina-equivalent rocks elsewhere. Isopach lines dashed where reconstructed across outcrop area. Map contrasts greatly with "Cayugan" thickness maps based on distribution of Salina Group proper.

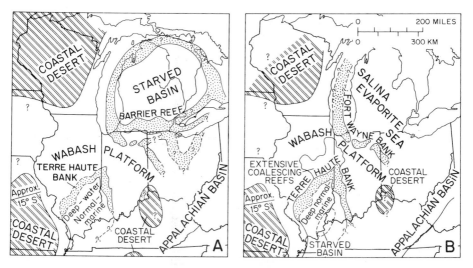

FIG. 7—Map showing general Silurian paleogeography, Midwest region. (*A*) Late Wenlockian and early Ludlovian; (*B*) Pridolian.

Formation, and the Louisville Limestone in the area south of the Fort Wayne bank. This sequence, including the part designated simply as lower Salina in northern Indiana, is a facies of A-carbonates and A-evaporites in Michigan, and of the Greenfield Dolomite and part or all the Tymochtee Formation (Salina Group) of western Ohio. It represents a restrictive influence on the platform and shows cyclicity. A typical cyclic deposit is 10 to 20 ft (3 to 6 m) thick and is composed of a lower dark, locally cherty, micritic carbonate and an upper light-colored, coarser grained carbonate representing, respectively, restricted and less-restricted environments.

Another, younger conspicuous Salinan sequence that extends both through and over the Fort Wayne bank and onto the platform had long been known as the Kokomo and Kenneth Limestones (Fig. 1). The lower, laminated micritic Kokomo Member represents return of restrictive conditions to the central platform during middle (possibly D-unit) Salinan deposition, and after an intervening period of more normal salinity, represented by the Mississinewa and by C-unit rocks, at least in part (Figs. 1, 8). The upper, Kenneth part of this Salinan sequence includes reeflike buildups, and is evidence of return to more normal-marine conditions on the Wabash platform during later Cayugan time (as late as E- or F-unit deposition?).

Terrigenous clastics from a southerly source are prominent in the Waldron and Mississinewa and help in correlation of salinity cycles. Some of the clastics entered distal parts of the protobasins as Waldron clastics in the Illinois basin, and as part of the Salina A-carbonates in the Michigan basin. Clastics of the Mississinewa are recognized in the Moccasin Springs Formation of the Illinois basin and possibly as parts of shales ranging from A to C in the Michigan basin.

Clastics entered the basin through passes in the barriers. Also, environmentally restrictive brines were interchanged between the Michigan basin and the platform. In environmentally favorable locales, reefs flourished, so that sequences of the platform

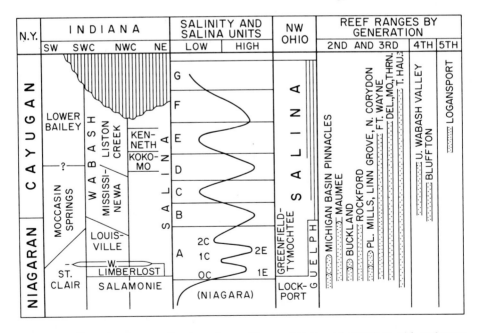

FIG. 8—Chart showing coordination of exemplary reef starts and abortions with carbonate-evaporite cycles for Michigan basin and Wabash platform (see Figure 1 for such data on the entire Great Lakes area). Open tops on range bars indicate eroded tops. Symbols: W. - Waldron; DEL. - Delphi; MO. - Monon; T. HAU. - Terre Haute; THRN. - Thornton.

and basin-margin actually are complexly interleaved facies of Salina-like micrites, more normal-marine, partly terrigenous, clastic-bearing interreef strata, and reef and reef-detrital bodies. Because of passes in the bank in southern Michigan, and in the Fort Wayne bank, the facies are complex (Fig. 5).

The lower Salina-like sequence (from top of Salamonie into Louisville [Fig. 1]) on the platform should represent the great restriction(s) or drawdown(s) that have been interpreted for the time of early deposition of the Salina in Michigan. Magnitudes of such events are debatable. Considering that southerly derived sediments twice pervaded the main study area, and that only transitional boundaries and no significant unconformities are recognized between or within the reef-bearing formations of the platform, history of the platform hardly appears to be compatible with the proposition of 400-ft (120 m) drawdown in the Michigan basin, development of major unconformities, and sabkhas as the principal evaporite-forming environments. Much greater compatibility is found in the principle of evaporite deposition advocated by Matthews and Egleson (1973). They believed that the depositional edge of a Salinan evaporite facies does not define a shoreline, but rather the extent of a transgressing and regressing body of brine acting on the seafloor. A shrinking sea was not necessary, and water depths providing for basin reflux could have been maintained at all or most times over the basin rims and through reef tracts. Salt is absent from Indiana; instead, the section is carbonate rock representing in part restricted environments that advanced far ahead of the salt, and onto the platform.

These interpretations appear to be in accord with the principal reef starts and abortions in the Wabash platform area. A first, partly aborted generation on the platform (second generation, Figs. 1, 5) is represented by many outcropping reefs near the Indiana-Ohio border (Fig. 4; Shaver, 1974b; Droste and Shaver, 1976; Indiana University Paleontology Seminar, 1976). These reefs, mostly less than 50 ft (15 m) thick, represent Niagaran time and the time of early Salinan A-unit deposition. They record a scaled-down model of, and represent imperfectly the same events as the entire sequence of pinnacle reefs in the Michigan basin, which is hundreds of feet thick. Some of these reefs show multiple starts (including 3rd generation, Figs. 1, 5) and algal-stromatolite caps, much like pinnacle reefs of the basin. Reefs of the second generation that were situated more favorably with respect to topography, passes in the barrier(s), or brine fronts include those at Montpelier, Yorktown, and Lapel (Fig. 4; Cumings and Shrock, 1928; Shaver, 1974a; Wahlman, 1974), but erosion has reduced their tops to modest levels in the overall stratigraphic section. The largest outcropping, partly eroded reefs of this same generation, some thicker than 400 ft (120 m), are in northwestern Indiana and northeastern Illinois. Many more, partly coalesced reefs of this generation exist southward in the subsurface of Indiana and Illinois.

The fourth generation of reefs of Figures 1 and 5 is represented on the Wabash platform by the group of erosion-etched reefs in the upper Wabash valley (Georgetown to Bluffton, Fig. 4; Cumings and Shrock, 1928; Textoris and Carozzi, 1964; Suchomel, 1975; Sunderman and Mathews, 1975). Several corings and quarry exposures now prove bottoming-out of these reefs in the position of the lower Mississinewa to upper Louisville. Their start thus was associated with the ending of the restricted environment represented by the underlying dense, unfossiliferous carbonate rocks of the Louisville (Salinan A-unit equivalent). None of this generation is known to have been aborted within the reef-bearing section, which could have lost through erosion minimally 200 ft (61 m) of section in the upper Wabash valley.

The numbered reef scheme devised here for the Great Lakes area might break down in platform locales most distant from and least associated with the cyclic carbonate-evaporite regime of the proto-Michigan and Appalachian basins. Nevertheless, some coordination may have been effected by interregional control of an epeirogenic kind (and/or eustatic sea-level changes) that is suggested by the periodic derivation and transport of terrigenous clastics; this control could have affected or effected the carbonate-evaporite cyclicity. A principal generation of Moccasin Springs reefs in Illinois (Lowenstam, 1949; Rogers, 1972), for example, seems to be unrelated to any strictly

Michigan-based cyclicity. However, they seem to represent the same generation as reefs of the upper Wabash valley (fourth generation, Indiana column, Fig. 1).

The last reef generation (fifth generation, Figs. 1, 5) consists of small eroded bio-stromes or buildups (tens of feet thick) in the Kenneth Limestone Member. In their particular locales, Logansport and southward in the structurally low area separating the Cincinnati and Kankakee arches (Figs. 2, 4), they began to develop when the restrictive depositional environment of the Kokomo Limestone Member ended (Fig. 8).

General Paleogeography

During middle Wenlockian to early Ludlovian time (middle to late Niagaran), the Terre Haute bank flanking the proto-Illinois basin (Vincennes basin of Droste and Shaver, 1975) had formed; so had the inner barrier-and-pinnacle reef system of the proto-Michigan basin, which afforded basin-marginal relief as part of the Salinan A-unit evaporite regime (Fig. 7A). Only the rudiments of a fairly open Fort Wayne bank had formed during this time, so that evaporite-related brines had access to the northern Wabash platform area and there adversely affected growth of some individual reefs.

During Pridolian (middle and late Cayugan) time (Fig. 7B), the Terre Haute bank reached a large size and included a northern extension along the Indiana-Illinois

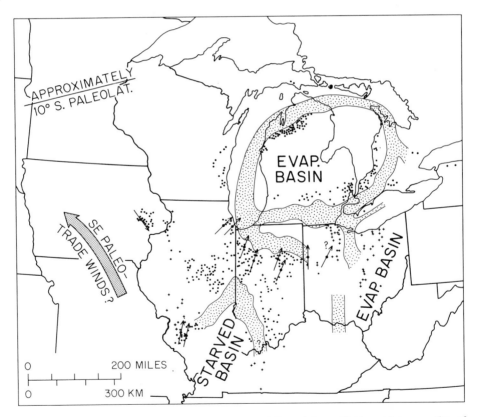

FIG. 9—Map of Midwest showing composite Middle and Late Silurian paleogeography of Silurian reef archipelago and fore-reef to back-reef directions (arrows) for all reefs known to have such directions reported: Iowa-Palisades; Illinois-Marine, Bartelso, Lemont, and Thornton; Indiana-Monon, Delphi, Wabash, and Montpelier; Ohio-Rockford and Buckland. Reeflike barriers shown by stippled pattern; discrete reefs (not all shown) by dots and stars (questionable status).

border in the form of large coalesced reefs; the inner reef system of the Michigan ¹ ısin had ceased to exist as an effective system. However, a mostly younger Fort Wayne bank marked a younger, southern margin of the basin, which, during middle and late deposition of the Salina Formation, continued to have some relief and bordered the evaporite basin to the north. The western basinal margin may have been more stable than the southern margin, as tentatively shown by Figures 7A and 7B, but Silurian rocks in eastern Wisconsin should be evaluated in these terms—if erosion has not removed most of the needed evidence.

The Midwest presumably was in low southern latitudes during Silurian time, and perhaps was in what could have been the belt of southeast tradewinds (Fig. 9). On the basis of present global winds, the assumed paleowind direction has a perhaps-unexpected vector relation with the fore-reef to back-reef directions reported for several platform reefs. One causative factor could be drift of water toward the Michigan evaporite basin that Berry and Boucot (1970, p. 88-89) invoked to explain the distributional pattern of limestone and dolomite suites among Silurian reef and interreef strata, including evaporites. Difficulties arise with any overriding single explanation of fore-reef directions, but the idea expressed by Berry and Boucot lends support to our proposal for contemporaneity of the grossly considered Silurian reef and evaporite episodes. Given the great geographic latitude available for its operation, overall contemporaneity should not be a difficult concept, whatever the relatively minor differentiation that applies to the reefs and evaporites that are in close proximity.

Conclusion

The multistage pinnacle reefs of the Michigan basin were short-lived and compare episodically with much smaller (where aborted) and partly multistage reefs on the basin-platform margin in Indiana and Ohio. Growth and abortion of these reefs was related to Salina A-unit (late Niagaran) cyclicity. The multistage Maumee, Ohio, reef probably survived longer (Fig. 8; Janssens, 1974, p. 85-87). Many other reefs of the same generation—possibly including parts of the Fort Wayne bank and other presently eroded reefs—could have survived actively until the end of Silurian time. Later periods of reef genesis were coordinated with salinity cycles associated with middle and late deposition of the Salina (after deposition of unit D).

Conceivably, the proto-Michigan basin could have been emergent periodically, around its margins in particular, thus explaining the facts interpreted as evidence of sabkhas, and it could have been basinlike in some areas, in the sense of rates of subsidence but not necessarily in depths of water. Nevertheless, the platform and basin-margin facies support the ideas of a rather steady state but with cyclic restriction of the basin, of water over the basin rims at most or all times, and of deeper-water deposition of thick evaporites. Further, interpretations of karst and freshwater pisolitization in reefs, which call for hundreds of feet of vertical exposure, should be re-examined. The genetic and diagenetic phenomena described for carbonate rocks by Scholle and Kinsman (1974) and by Estaban and Pray (1975), for example, should be considered.

The interrelated structural and sedimentational developments of the Wabash platform and bordering protobasins were complex during Silurian time. Therefore, the search for reef-associated hydrocarbons is made both more difficult and attractive. It is more difficult because there is no single reef generation of one stratigraphic definition, and there is no single geographically definable reef belt for each basin; it is more attractive because the observed complexity suggests large, virtually unexplored areas and sections of rock.

References Cited

Alling, H. R., and L. I. Briggs, 1961, Stratigraphy of the Upper Silurian Cayugan evaporites: AAPG Bull., v. 45, p. 515-547.

Berry, W. B. N., and A. J. Boucot, 1970, Correlation of the North American Silurian rocks: Geol. Soc. America Spec. Paper 102, 289 p.

Collinson, Charles, et al, 1967, Devonian of the north-central region, United States, *in* D. H. Oswald, ed., International symposium of the Devonian System: Alberta Soc. Petroleum Geologists, v. 1, p. 933-971.

Crowley, D. J., 1973, Middle Silurian patch reefs in Gasport Member (Lockport Formation), New York: AAPG Bull., v. 57, p. 283-300.

Cumings, E. R., and R. R. Shrock, 1928, The geology of the Silurian rocks of northern Indiana: Indiana Dept. Conserv. Pub. 75, 226 p.

Droste, J. B., and R. H. Shaver, 1975, Jeffersonville Limestone (Middle Devonian) of Indiana: stratigraphy, sedimentation, and relation to Silurian reef-bearing rocks: AAPG Bull., v. 59, p. 393-412.

—— —— 1976, The Limberlost Dolomite of Indiana: a key to the great Silurian facies in the southern Great Lakes area: Indiana Geol. Survey Occasional Paper 15, 21 p.

—— —— J. D. Lazor, 1975, Middle Devonian paleogeography of the Wabash platform, Indiana, Illinois, and Ohio: Geology, v. 3, p. 269-272.

Estaban, Mateo, and L. C. Pray, 1975, Subaqueous, syndepositional growth of in-place pisolite, Capitan reef complex (Permian), Guadalupe Mountains, New Mexico and West Texas (abs.): Geol. Soc. America Abs. with Programs, v. 7, p. 1068-1069.

Felber, B. E., 1964, Silurian reefs of southeastern Michigan: Unpub. Ph.D. thesis, Northwestern Univ., 104 p.

Fincham, W. J., 1975, The Salina Group of the southern part of the Michigan basin: Unpub. M.S. thesis, Michigan State Univ., 56 p.

—— J. H. Fisher, 1975, Salina Group in southern Michigan basin (abs.): AAPG Bull., v. 59, p. 1736.

Fisher, J. H., 1973, Petroleum occurrence in the Silurian reefs of Michigan: Ontario Petroleum Inst., 12th Ann. Conf., Paper 9, 10 p.

Gill, Dan, 1975, Cyclic deposition of Silurian carbonates and evaporites in Michigan basin—discussion: AAPG Bull., v. 59, p. 535-538.

Grabau, A. W., 1909, Physical and faunal evolution of North America during Ordovic, Siluric, and early Devonic time: Jour. Geology, v. 17, p. 209-252.

Indiana University Paleontology Seminar, 1976, Silurian reef complex at Rockford, Ohio; constitution, growth, and significance: AAPG Bull., v. 60, p. 428-451.

Janssens, Arie, 1974, The evidence for Lockport-Salina facies changes in the subsurface of northwestern Ohio: Michigan Basin Geol. Soc. Ann. Field Conf., p. 79-88.

Liberty, B. A., and T. E. Bolton, 1956, Early Silurian stratigraphy of Ontario, Canada: AAPG Bull., v. 40, p. 162-173.

Lowenstam, H. A., 1949, Niagaran reefs in Illinois and their relation to oil accumulation: Illinois Geol. Survey Rept. Inv. 145, 36 p.

—— 1950, Niagaran reefs of the Great Lakes area: Jour. Geology, v. 58, p. 430-487.

Matthews, R. D., and G. C. Egleson, 1973, The origin and implications of a mid-basin potash facies in the Salina salt of Michigan: Northern Ohio Geol. Soc., 68 p. (preprint).

Mesolella, K. J., et al, 1974, Cyclic deposition of Silurian carbonates and evaporites in Michigan basin: AAPG Bull., v. 58, p. 34-62.

—— —— 1975, Cyclic deposition of Silurian carbonates and evaporites in Michigan basin—reply: AAPG Bull., v. 59, p. 538-542.

Okla, Saleh, 1976, Subsurface stratigraphy and sedimentation of Middle and Upper Silurian rocks of northern Indiana: Unpub. Ph.D. thesis, Indiana Univ., 143 p.

Pollock, C. A., and C. B. Rexroad, 1973, Conodonts from the Salina Formation and the upper part of the Wabash Formation (Silurian) in north-central Indiana: Geologica et Palaontologica, v. 7, p. 77-92.

Pounder, J. A., 1963a, Guelph-Lockport drilling should reveal more reefs: Oil and Gas Jour., v. 61, p. 144-148.

—— 1963b, Structure, economics play key roles in Guelph-Lockport search: Oil and Gas Jour., v. 61, p. 162-164.

Rickard, L. V., 1975, Correlation of the Silurian and Devonian rocks in New York State: New York State Mus. and Sci. Service Map and Chart Ser. No. 24, 16 p.

Rogers, J. E., Jr., 1972, Silurian and Lower Devonian stratigraphy and paleobasin development; Illinois basin - central United States: Unpub. Ph.D. thesis, Univ. Illinois, 144 p.

Ross, C. A., 1962, Silurian monograptids from Illinois: Palaeontology, v. 5, p. 59-72.

Scholle, P. A., and D. J. J. Kinsman, 1974, Aragonitic and high-Mg calcite caliche from the Persian Gulf—a modern analog for the Permian of Texas and New Mexico: Jour. Sed. Petrology, v. 44, p. 904-916.

Schuchert, Charles, 1943, Stratigraphy of the eastern and central United States: New York, John Wiley & Sons, 1,013 p.

Sharma, G. D., 1966, Geology of Peters reef, St. Clair County, Michigan: AAPG Bull., v. 50, p. 327-350.

Shaver, R. H., 1974a, Silurian reefs of northern Indiana—reef and interreef macrofaunas: AAPG Bull., v. 58, p. 934-956.

—— 1974b, Structural evolution of northern Indiana during Silurian time: Michigan Basin Geol. Soc. Field Conf., p. 55-77, 89-97, 102-111.

—— et al, 1971, Silurian and Middle Devonian stratigraphy of the Michigan basin—a view from the southwest flank: Michigan Basin Geol. Soc. Ann. Field Conf., p. 37-59.

Sloss, L. L., 1947, Environments of limestone deposition: Jour. Sed. Petrology, v. 17, p. 109-113.

Suchomel, Diane, 1975, Paleoecology and petrology of Pipe Creek, Jr. reef (Niagaran-Cayugan), Grant County, Indiana: Unpub. A.M. thesis, Indiana Univ., 38 p.

Sunderman, J. A., and G. W. Mathews, eds., 1975, Silurian reef and interreef environments: Fort Wayne, Indiana Univ. - Purdue Univ., 94 p.

Textoris, D. A., and A. V. Carozzi, 1964, Petrography and evolution of Niagaran (Silurian) reefs, Indiana: AAPG Bull., v. 48, p. 397-426.

Wahlman, G. P., 1974, Stratigraphy, structure, paleontology, and paleoecology of the Silurian reef at Montpelier, Indiana: Unpub. A.M. thesis, Indiana Univ., 71 p.

Williams, M. Y., 1919, The Silurian geology and faunas of Ontario Peninsula, and Manitoulin and adjacent islands: Canada Geol. Survey Mem. 111, 195 p.

Wood, G. D., 1975, Acritarchs and trilete spores from the Mississinewa Shale of northern Indiana, in J. A. Sunderman and G. W. Mathews, eds., Silurian reef and interreef environments: Fort Wayne, Indiana Univ. - Purdue Univ., p. 91-94.

Zenger, D. H., 1965, Stratigraphy of the Lockport Formation (Middle Silurian) in New York State: New York State Mus. and Sci. Service Bull. 404, 210 p.

Seismic Data-Processing Techniques in Exploration for Reefs, Northern Michigan[1]

P. L. MCCLINTOCK[2]

Abstract Difficulty in detecting reefs in northern Michigan is caused mainly by noise, "multiples" (reverberations), variations in topography, and variations in thickness of weathered material and glacial drift. During the last three years new data-processing techniques—such as automatic statics and deconvolution, and relative-amplitude processing methods—have led to improved detection of reefs. Quality of information from seismic sections is enhanced further as data-processing geophysicists gain experience in the areas of activity. Communication among data-processing geophysicists, interpreters, and geologists also is necessary to produce the most reliable picture of subsurface geology in the reef tracts.

Introduction

Seismic data recordings contain valid reflection information and erroneous data that commonly often mask the legitimate reflection information. The major problems with seismic information in northern Michigan are noise (both random and coherent), "multiples" or reverberations, and weathering and drift variations.

Noise and Multiples

Seismic records contain noise of two types—random noise and coherent noise (Fig. 1). Random noise, which generally can be reduced by using proper bandpass filters during processing, is not considered to be a major problem. Coherent noise—and in particular, ground roll—interferes with and commonly masks the primary reflection information. Thus, coherent noise it is a major problem. Efforts are made during field acquisition to cancel coherent noise through proper source and detector arrays, but the noise energy is so strong that it never can be completely canceled. In northern Michigan separation of coherent noise and reflection energy is extremely difficult because frequency ranges of coherent noise and reflections are the same. Therefore, use of bandpass filters does not eliminate coherent noise.

An effective solution to problems of coherent noise has been to apply deconvolution, in conjunction with common-depth point (CDP) stacking. The deconvolution process, performed on raw data before CDP stacking, attenuates or removes part of the repetitive energy from coherent noise.

Seismic multiples are analogous to ordinary echoes; the primary reflection is similar to the first echo that one hears, and the multiples are similar to the echoes that follow. Deconvolution removes the secondary echoes from seismic records. In northern Michigan, multiples are short-period and probably are induced at the contact of glacial drift and bedrock.

Following deconvolution, subsurface positions are resampled with different source-detector spacings. Because interference arrivals are related directly to source-detector distance, summation of a large number of common-depth point signals that have various source-detector spacings gives a large reduction in coherent noise, as well as random noise.

Variations in Surficial Materials

Variation of topography, and of thickness of weathered material and glacial drift, causes the largest problem in obtaining a reliable seismic section. These factors change continuously along seismic lines and introduce time shifts in the data that misalign

[1] Manuscript received January 5, 1976; accepted August 17, 1976.

[2] Seismograph Service Corporation, Tulsa, Oklahoma (now with Western Geophysical, Houston, Texas, 77001.

The writer thanks David Hall, President, Reef Petroleum Corporation, and Randy Ingalls, Northern Michigan Exploration Company, for permission to use the data shown in this paper.

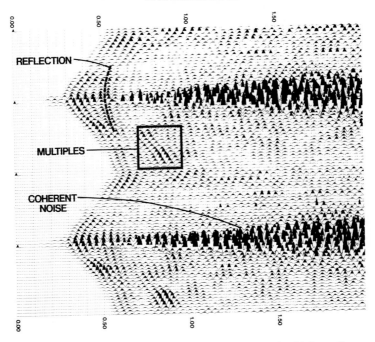

FIG. 1–Seismic record showing examples of actual reflections, and misinformation as multiples and coherent noise. Horizontal scale in seconds.

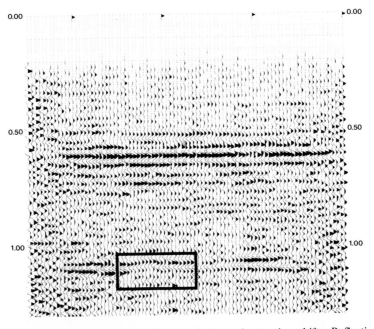

FIG. 2–Seismic record that shows misaligned reflections due to time shifts. Reflections (inset) could be misinterpreted as revealing a reef. Vertical scale, seconds.

reflections. If these time shifts are not accurately determined, the section may be misinterpreted. Figure 2 shows an example in which part of the record indicates presence of a reef; the data are misleading, owing to time shifts. When the shifts are determined and data are corrected (Fig. 3) a reliable interpretation can be made, demonstrating that no reef is detectable from the record.

Determination of time shifts or "statics" is difficult and time-consuming. Commonly, several iterations of velocity corrections, automatic statics, and hand-picked statics are necessary to solve this problem.

The degree of success a data-processing geophysicist has in removing unwanted data depends on how well he chooses the parameters for elimination. There are instances, of course, where the unwanted data are of such magnitude that the real reflection information is virtually impossible to extract. However, important strides in recording capability and field techniques have improved the basic quality of data. Specific discussion of these changes is outside the purpose of this paper, but such improvements have contributed to the improved success ratio in detection of reefs in northern Michigan.

Data Processing

The main goal of seismic data processing is to produce a valid picture of the subsurface. The role of each processing step is to remove extraneous data from records. If optimum data processing techniques could be designed, only real reflection information would remain. Recent improvements include relative-amplitude processing, seismic modeling and well synthetics. These and other techniques contribute to improved interpretations and data quality; they have become a major part of the overall data-processing procedure.

Seismic problems can be identified and listed, but there are no universal solutions to them. Solutions that work well on one line often do not work on a line a mile away. Therefore, each line must be studied carefully by the processing geophysicist. His ability to recognize and evaluate properly extraneous data on the records is dependent on his knowledge of the area and his experience with data from it.

A very important part of the continuing education of the processing geophysicist rests with oil-company geologists. In order to insure that data are processed properly, geologists should instruct the processing geophysicist in the geology of the area, in details of the strata of interest, goals to be accomplished, information from wells in the area, and so forth. Even after a well is drilled, results should be discussed with the processing geophysicist, particularly in the case of a dry hole. Through this feedback, data processors can improve their ability to solve seismic problems and to help interpreters improve success ratios in the search for oil and gas.

Figure 4 is a simplified flow chart that demonstrates some of the processing steps that go into a final seismic section. Each step requires analytical decisions by the geophysicist, which demonstrates that while seismic data processing has become computer-dependent for repetitive calculations, the analyst remains the key decision-maker in determining the quality of the final section.

Example From Manistee County, Michigan

A seismic line from Manistee County, Michigan (Fig. 5) was chosen to demonstrate how a processing geophysicist would attempt to solve problems associated with records from the area. The line is located in Secs. 20 and 29, T23N, R15W, Bear Lake Township (Fig. 6). It was recorded digitally with binary gain on August 30, 1974, using a dynamite energy source, 48-trace recording, and 2,400% subsurface coverage.

The line was originally processed in September, 1974, and reprocessed in August, 1975. Procedures used in reprocessing were the result of techniques refined as some of the local geophysical problems were solved, and more information became available about the geology of the area.

No attempt will be made to describe rigorously the details of each step of analysis; rather, the discussion will show how assumptions made by the processor affected the

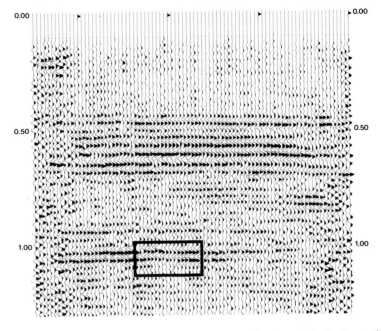

FIG. 3—Seismic record shown on Figure 2, with data corrected for time shifts. Vertical scale, seconds.

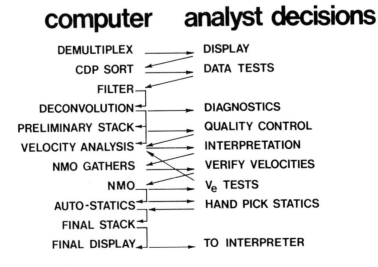

FIG. 4—Simplified flow chart showing steps in analysis of a seismic section.

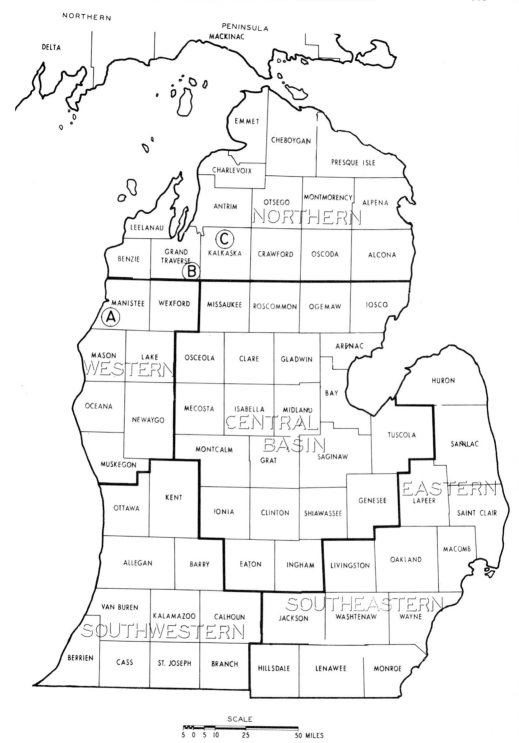

FIG. 5–Locations of Manistee (A), Grand Traverse (B), and Kalkaska (C) counties, Michigan. Seismic lines used as examples in this paper were recorded in these counties.

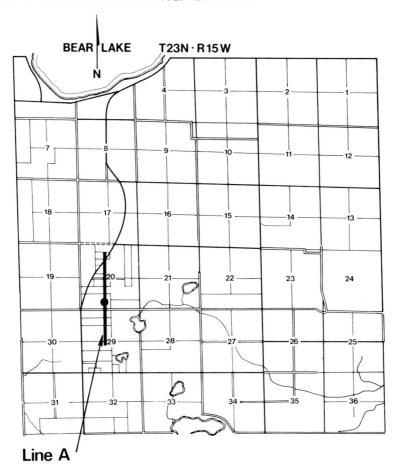

FIG. 6–Location of seismic line, used to show effectiveness of improved analytical techniques.
Area shown is in Manistee County, Michigan.

final section. Progressive improvement at the Niagaran level, around 0.75 sec, is noteworthy.

It should be pointed out that nearly every parameter used in seismic data processing
is affected by the geologic conditions of the area and of the zones of interest. Therefore, in advance of actual processing, the geophysicist reviewed well logs, formation
tops, completion reports and other seismic data gathered in the area. He also sought
geologic and seismic information about the area from geologists and the interpreter.
(In new areas, however, seismic models that incorporate known geologic conditions
are made as a basis for hypotheses about how the seismic section should relate to the
local geology.)

Next, individual records were examined. From these, the processing geophysicist
determined that the data were recorded properly, determined parameters for gain
recovery, identified severe geophysical problems associated with the data, and formu-
lated a plan to attenuate noise and multiples. Data were sorted into common-depth
point order to group those traces that represent a subsurface point. In northern Michigan, there normally is a redundancy of 24 traces for each common-depth point. This is
so that random noise will be partially canceled when the data are stacked, improving

Line A
PRELIMINARY STACK

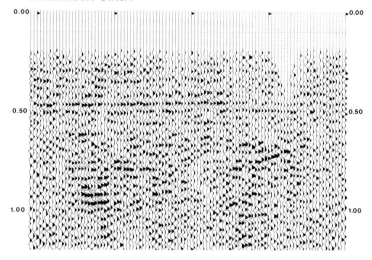

FIG. 7–Stacked seismic section, Secs. 20 and 29, T23N, R15W, Manistee County, Michigan. (In illustrations 7-11, improvement of information at Niagaran level [about 0.75 sec] is noteworthy.)

the signal-to-noise ratio of the final section. Additionally, corrections for variations of elevation were computed and applied to the data.

Tests on the data, such as autocorrelations to determine deconvolution parameters, and frequency and filter comparisons to attenuate random noise, helped the data processor determine optimum parameters for filtering random noise and eliminating multiples. After deconvolution and filtering, additional testing verified the effectiveness of these processes.

Velocities were determined by use of constant velocity stacks, whereby the time and velocity that produced the best stack were formed into a velocity function. This function was checked for validity by comparison of it with well logs and other pertinent information from the area. The velocity function was then applied to the data to produce the first stacked section (Fig. 7).

Elevation variations were corrected using a velocity of 3,000 ft/sec (910 m/sec). This correction was based on experience with the near-surface section from seismic lines nearby and from tests using different correction velocities. The Devonian reflection at 0.46 sec is nearly flat across the section (Fig. 7).

After deconvolution, filtering, and stacking (Fig. 8), the Devonian was more apparent than on the preliminary stack (Fig. 7), but the Niagaran reflection at 0.76 sec was still not completely visible.

Originally, the line had elevation corrections computed using a velocity of 6,000 ft/sec (1,829 m/sec) (Fig. 9) which made the Devonian reflection more erratic, particularly on the right-hand two-thirds of the section. This deterioration is most evident in the interval from 0.45 to 0.60 sec, and corresponds to elevation variations of about 40 ft (12 m) across the line. After the best choice of elevation corrections was determined, the velocity function was recomputed and checked to verify its validity.

Automatic statics (Fig. 10) was the next step performed to reduce static errors. Notice the Niagaran reflection now is quite visible on the section. Next, hand-picked statics were derived and automatic statics re-run to yield the final section (Fig. 11). On many lines, many trials of hand-picked and automatic statics must be performed to correct the data fully.

Line A
V_e=3000

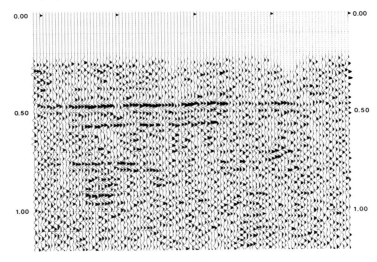

FIG. 8–Seismic line shown on Figure 7, after deconvolution, filtering, and stacking. Note improved visibility of Devonian reflection at about 0.46 sec.

Line A
V_e=6000

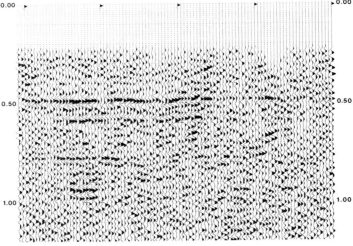

FIG. 9–Seismic line (shown on Figs. 7, 8) including elevation corrections, computed with velocity of 6,000 ft/sec (1,830 m/sec).

Line A
AUTOMATIC STATICS

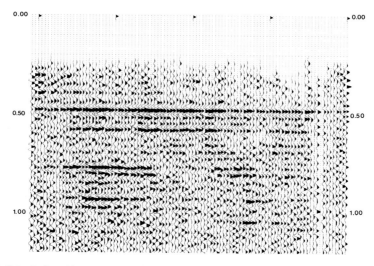

FIG. 10—Seismic line (shown in Figs. 7, 8, 9) including automatic-statics correction. Reflection of Niagaran strata (about 0.75 sec) is improved.

Line A
FINAL STACK

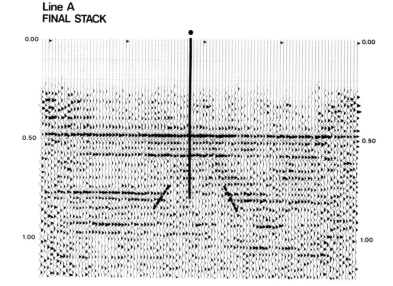

FIG. 11—Seismic line (shown on Figs. 7 through 10). Interpretation shows location of Niagaran reef at about 0.75 to 0.85 sec, and approximate location of discovery well.

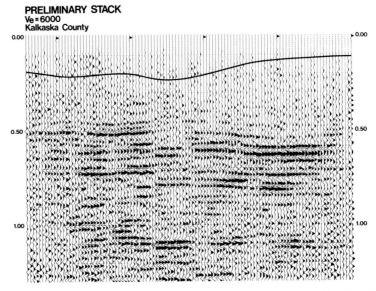

FIG. 12–Seismic line, Kalkaska County, Michigan. Solid line is elevation profile. Note that reflection at approximately 0.5 sec has form similar to elevation profile.

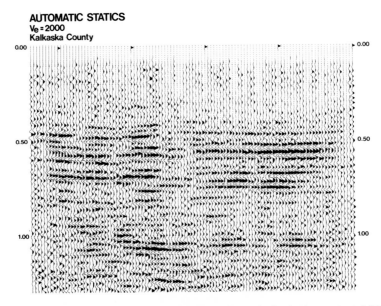

FIG. 13–Seismic section, Kalkaska County, Michigan. Record of seismic event at 1.05 sec was improved by hand-statics correction (cf. Figs. 12, 13).

FINAL STACK
Kalkaska County

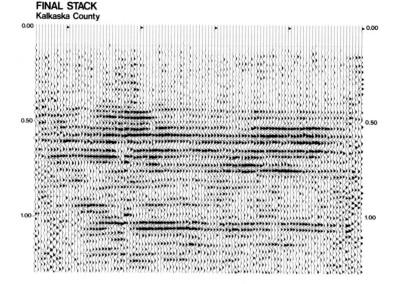

FIG. 14—Seismic section, Kalkaska County, Michigan. Seismic event at 1.05 sec shows improved record after hand-statics correction (cf. Fig. 13).

Finally, sections were reviewed with the interpreter to be sure that the final section was a reasonable picture of the subsurface geology, as based on the interpreter's knowledge of the area. As a result of the interpretation made on this line, Reef Petroleum Corporation drilled the Miller 1.20 well which flowed 650 bbl (103 cu m) of oil and 560,000 cu ft (15,850 cu m) of gas per day.

Examples From Kalkaska and Grand Traverse Counties, Michigan

Three lines are shown that demonstrate problems with noise, multiples, and weathering variations. The line from Kalkaska County (Fig. 12) illustrates the false structures caused by variation of the stratigraphic section above bedrock. On this line, both the thickness and velocity of surficial materials vary (Fig. 12). This section was deconvolved and filtered prior to display. Test-stacking the data with different elevation-correction velocities indicated that 2,000 ft/sec (610 m/sec) corrected most of the line. The erroneous data probably were caused by loose, sandy materials above bedrock.

However, a statics problem still remained near the center of the section, even after running automatic statics (Fig. 13). Therefore, hand statics were picked to correct the remaining velocity and thickness variations. The final section (Fig. 14) shows the effectiveness of the combined automatic and hand statics. Note the improved continuity of the event at 1.05 sec (Fig. 14) as compared to the apparent anomaly shown in Fig. 13. Obviously, an erroneous interpretation could have been made if the problem of statics had not been corrected.

The preliminary section of a seismic line from Grand Traverse County (Fig. 15) demonstrates how noise and amplitude variation make the section impossible to interpret, particularly at the level of Niagaran strata. After having been fully processed, the final section (Fig. 16) was interpretable, and an anomaly was identified that led to discovery of commercial production.

A third seismic line (Fig. 17), also from Kalkaska County, demonstrates how noise and multiples distort or obliterate reflection events. After deconvolution, filtering, and automatic statics, the final section (Fig. 18) showed an anomaly—a reef that was drilled. Note the improved signal at the Niagaran level and attenuation over the reef.

PRELIMINARY STACK
Grand Traverse County

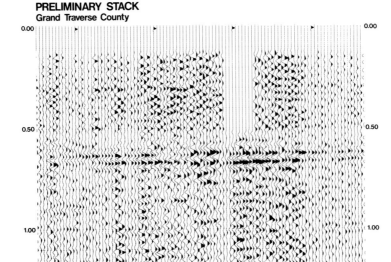

FIG. 15–Preliminary seismic line, Grand Traverse County, Michigan. Section of Niagaran strata at about 1.0 sec is difficult to interpret.

FINAL STACK
Grand Traverse County

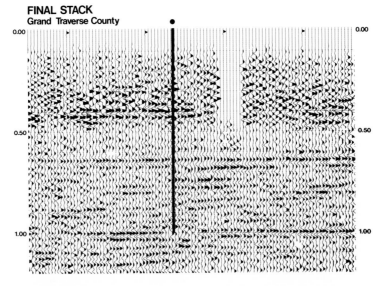

FIG. 16–Processed seismic line shown on Figure 15. Record of Niagaran strata shows improvement; anomaly indicated by solid vertical line is an oil field. (Scale, seconds).

PRELIMINARY STACK
Kalkaska County

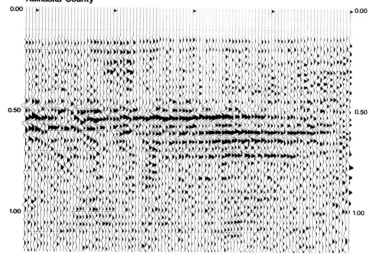

FIG. 17—Preliminary seismic line, Kalkaska County, Michigan. Compare signal at level of Niagaran strata (1.0 to 1.1 sec) with signal in final section (Fig. 18).

FINAL STACK
Kalkaska County

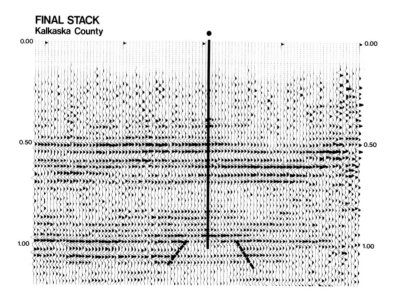

FIG. 18—Processed seismic line from data shown on Figure 17. Note increased resolution of Niagaran section. (Scale, seconds).

Obviously, the data processor must know a great deal about the data on which he is working. Through the feedback he receives from geologists and interpreters, the data processing geophysicist is able to increase his knowledge, improve his processing ability, and improve the reliability of the final section.

Problems Remaining

Although many advances have been made, and success ratios are improving, a great deal of work remains to solve fully the problems of seismic data processing. For instance, better techniques for automatic statics are needed, especially in instances where coherent-noise levels are high. Static corrections for varied thicknesses of weathered-zones and glacial drift are the largest problem yet to be overcome.

Additional research into relative amplitude and frequency variations is promising and could help differentiate between real and false anomalies. The increased use of seismic modeling helps to improve geophysicists' knowledge of the geology and seismic appearance of reefs.

Conclusions

The difficulty of detecting reefs in northern Michigan is caused mainly by noise, multiples, and variations in surficial materials. Techniques to overcome these problems have improved steadily as computer programs are developed and data-processing geophysicists gain more experience and knowledge.

As was demonstrated in the examples set out above, seismic data processing involves many decisions and many processing steps. However, no matter how sophisticated the processing techniques become, no data-processing geophysicist can produce sections of the highest quality without intimate knowledge of the geology of the area. Therefore, cooperation among geophysicists, geologists, and interpreters is extremely important to production of the most reliable picture of the subsurface.

Relationships Between Depositional Environments, Tonoloway Limestone, and Distribution of Evaporites in the Salina Formation, West Virginia[1]

R. A. SMOSNA[2], D. G. PATCHEN[2], S. M. WARSHAUER[3], and W. J. PERRY, JR.[4]

Abstract The Upper Silurian Tonoloway Limestone at Pinto, Maryland, is divided into three informal members on the basis of field, paleontologic, and petrographic studies. The lower member is characterized by thin bedding, stromatolites, gypsum molds, intraclasts, and mud cracks. The rocks typically are laminated micrite and pelmicrite. In the field and in thin section, this member is similar to the upper one. Faunal diversity is extremely low in both. The middle member, composed of biopelsparite and biomicrite, is more fossiliferous than either of the other members and shows an extreme variability in the development of communities. This member can be traced along the outcrop from Pinto, Maryland, to Greenbrier County, West Virginia.

Under the eastern Appalachian Plateau, dolomite and anhydrite occur in the upper and lower parts of the Tonoloway, whereas the middle member is chiefly limestone. Farther west, the subsurface equivalent of the Tonoloway is the Salina Formation, which consists of light to dark-gray dolomite and anhydrite with minor green and gray shale. Several salt beds generally are developed. Most of these are in the Salina F unit (equivalent to the upper Tonoloway member), but in Marshall County, West Virginia, salts as low in the section as the D evaporite (lower Tonoloway) are also well developed. Total salt thickness exceeds 200 ft (61 m) in several wells; however, excessive thicknesses may be due to salt flowage in anticlines.

The upper and lower members of the Tonoloway in the outcrop belt were deposited on intertidal-supratidal mud flats. In these two members beneath the eastern Plateau, salinity apparently increased toward the depocenter, and halite precipitated within the deeper central region of the evaporite basin. The evaporite basin was closed, with periodic influxes of normal sea water. The middle member of the Tonoloway was deposited in environments where water depth fluctuated from intertidal to shallow subtidal to deeper subtidal. A widespread transgression led to the more normal marine conditions during deposition of the middle member. Effects of this transgression can be traced into the basin, where the calcareous middle member supplanted evaporites.

Introduction

On Late Silurian regional paleogeographic maps and stratigraphic cross sections, West Virginia is often shown in question or incomplete (see, for instance, Alling and Briggs, 1961). The purpose of our work was to unravel the geologic history of West Virginia during Cayugan (Late Silurian) time. Our approach to this problem began with a study of the paleoenvironments of the Tonoloway Limestone in the eastern outcrop belt of the central Appalachians (Fig. 1). Based on the conclusions of this study, we continued our interpretation of Cayugan basinal history with an examination of the subsurface stratigraphy of the Salina Formation to the west (Fig. 1). Very little data are available on the Salina—only sample descriptions and gamma-ray logs of very few, widely scattered wells; we had no cores, chemical analyses, or mine exposures. Therefore, in order to establish the ancient setting of this unit, we related environments of deposition of the Tonoloway Limestone to the stratigraphic and geographic distribution of evaporites in the Salina Formation. Final deductions were based on a combination of surface and subsurface stratigraphy, paleontology, and carbonate petrology. With paleoenvironments of the Salina and Tonoloway understood, a regional framework was established for this part of the Appalachian basin during Late Silurian time. From this model, areas of possible oil and gas accumulation were then outlined.

[1] Manuscript received February 23, 1976; accepted June 30, 1976.

[2] West Virginia Geological and Economic Survey, Box 879, Morgantown, West Virginia 26505.

[3] Department of Geology and Geography, West Virginia University, Morgantown, West Virginia 26506.

[4] U.S. Geological Survey, Reston, Virginia 22092. Approved for publication by the director.

This paper is published with permission of Dr. Robert B. Erwin, Director, West Virginia Geological and Economic Survey. Nomenclature in the paper follows that of the West Virginia Geological Survey; it does not necessarily conform to usage of the U.S. Geological Survey.

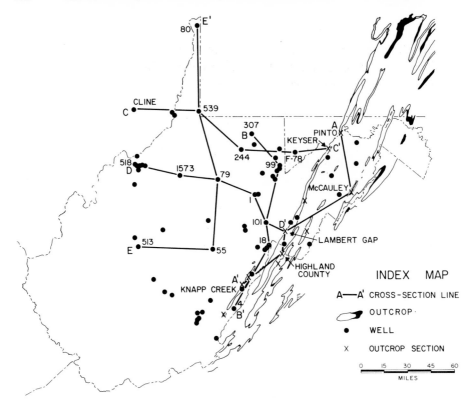

FIG. 1–Index map showing (1) sources of data and (2) locations of cross-section lines of Figs. 5 through 9, West Virginia and Maryland. Well numbers are state permit numbers.

The Tonoloway and Salina formations are of different lithologies—limestone on the one hand and evaporites on the other—and previously were never considered together in any stratigraphic scheme. As a matter of fact, previous investigations in West Virginia and adjacent areas have been limited (Swartz et al, 1923; Woodward, 1941; Martens, 1943; Cate, 1961, 1965; Ulteig, 1964; Fergusson and Prather, 1968; Rickard, 1969; and Clifford, 1973).

The stratigraphic interval of concern in this region (Fig. 2) is defined as follows: (1) The Tonoloway Limestone, basically laminated, fine-grained, sparsely fossiliferous, and somewhat argillaceous, is present in the outcrop belt and the eastern subsurface. (2) The Salina Formation of the western subsurface is defined as interbedded evaporites—anhydrite and salt—and dolomite, and correlates with the Salina D through G units of Ohio. Subsurface Salina units used in West Virginia follow those of Ulteig (1964) and Clifford (1973). (3) The immediate overlying formation is either the Keyser Limestone to the east, which is very fossiliferous, cherty, and knobby, or the Bass Islands Formation to the west, a shaly carbonate. (4) Underlying the interval are clastic units: the Wills Creek Formation, the Williamsport Sandstone, and the Salina C unit of Ohio.

Environments of Deposition of the Tonoloway Limestone

Field examination of eight Tonoloway sections (Fig. 1) together with detailed petrographic and paleontologic analyses of the classic Silurian section at Pinto, Maryland (Fig. 1), were used to interpret the paleoenvironments of the Tonoloway. On the basis of this study, three facies with distinct lithologies and characteristic fauna have

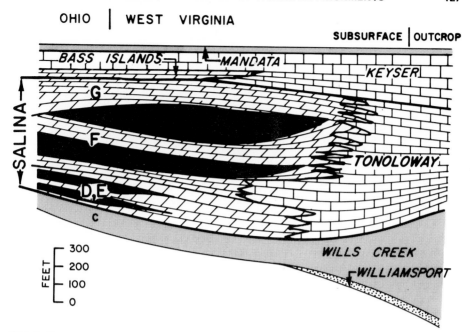

FIG. 2—East-west cross section of Upper Silurian - Lower Devonian strata of northern West Virginia. Stratigraphic nomenclature used in this report is shown. Salt is shown in black.

been noted (Fig. 3). These three environments of deposition are (1) the intertidal-supratidal mud flat, (2) shallow subtidal, and (3) deeper subtidal.

Intertidal-Supratidal Facies

The outcrop of the intertidal-supratidal facies displays such indicative features as laminated bedding, mud cracks, gypsum crystals and molds, halite-crystal impressions, and algal mounds. The algal structures are both thrombolites (Aitken, 1967) and LLH-type stromatolites (Logan et al, 1964), the latter being both closely spaced and widely spaced, single domes and complex domes, and ranging in diameter from 1 in. (2.54 cm) to more than 4 ft (1.2 m). Both kinds of algal structures thrived on protected intertidal flats (Logan et al, 1964; Aitken, 1967). The evaporite minerals resulted from subaerial diagenesis within either supratidal or intertidal sediments (Kinsman, 1969).

In thin section, these rocks are micrite and pelmicrite; only a trace of intramicrite was observed. Pellets are the most common allochems but even these are not abundant—most pelmicrites are mud-supported. Intraclasts generally have an internal texture of laminated micrite or pelmicrite; hence they represent desiccation chips of the tidal-flat sediments. Though microspar and sparry calcite bind the pellets of some laminae, micrite is dominant in this microfacies. Euhedral dolomite rhombs average about 25% in the samples. Small-scale scour, cut-and-fill, and unconformities are the results of minor current action. About half of the thin sections show burrowing, and many of these burrows are small and vertical, as to be expected within the intertidal zone (Rhoads, 1967).

Although these limestone beds are mostly unfossiliferous, two low-diversity communities, *Leperditia* and *Leperditia-Hormatoma*, are recognized. Thus it appears that only ostracods, gastropods, and blue-green   algae (stromatolites) were able to inhabit this harsh mud-flat environment. The great affinity displayed by the two invertebrate taxa for the stromatolites suggests that a feeding relationship may have existed, in which the two herbivores browsed on the blue-green algae.

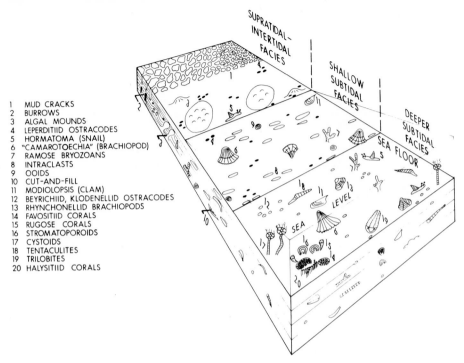

1 MUD CRACKS
2 BURROWS
3 ALGAL MOUNDS
4 LEPERDITIID OSTRACODES
5 HORMATOMA (SNAIL)
6 "CAMAROTOECHIA" (BRACHIOPOD)
7 RAMOSE BRYOZOANS
8 INTRACLASTS
9 OOIDS
10 CUT-AND-FILL
11 MODIOLOPSIS (CLAM)
12 BEYRICHIID, KLODENELLID OSTRACODES
13 RHYNCHONELLID BRACHIOPODS
14 FAVOSITIID CORALS
15 RUGOSE CORALS
16 STROMATOPOROIDS
17 CYSTOIDS
18 TENTACULITES
19 TRILOBITES
20 HALYSITIID CORALS

FIG. 3—Environments of deposition interpreted for the Tonoloway Limestone of the eastern outcrop belt.

Shallow Subtidal Facies

The shallow subtidal facies is distinguished by presence of intraclasts (often cherti-fied), edgewise conglomerates, and thin to medium bedding. These rocks are intra-sparite and intraclastic biopelsparite. The intraclasts and edgewise conglomerates repre-sent "rip-ups" from the intratidal-supratidal and deeper subtidal environments during storms and minor transgressions. Both true and superficial ooids are present in small amounts. Allochems are cemented by sparry calcite, that is, void-filling cement and overgrowths on fossils, yet micrite is invariably present to some degree as matrix be-tween grains, infilling of skeletons, or as interbedded micrite. Carbonate grains typical-ly are tightly packed and therefore support themselves. Minor scour along bedding planes was observed. Large-scale channels and burrows are conspicuously absent. Bed-ding is uniformly tabular.

Fossil diversity is moderate, and this microfacies is populated by a *Leperditia*-*"Camarotoechia"* community. The most common fossils include leperditiid and beyrichiocopid ostracodes, small globose brachiopods, cystoids, and pelecypods. Gastropods, calcareous algae, bryozoans, and tentaculitids and styliolinids are present in trace amounts. In general, the valved invertebrates are disarticulated, many of the fossils are fragmented (by scavengers and predators), and micritic envelopes (produced by boring algae) are widespread. Bioclasts are not rounded.

These intraclasts, as well as ooids and sparry calcite cement, all are evidence for a high-energy environment, but neither high enough to round fossils nor to wash away micrite completely. Rocks of this second facies have been interpreted to be shallow subtidal sediments on the basis of the moderate faunal diversity and the relatively high degree of agitation, plus the close association with intertidal-supratidal rocks.

Deeper Subtidal Facies

The deeper subtidal facies is indicated by abundantly fossiliferous limestone, consisting of biopelsparite, biopelmicrite and biomicrite. Two highly diverse communities are associated with these beds. The first is a strophomenid brachiopod (*Schuchertella?*) community composed of large numbers of brachiopods, bryozoans, palaeocopid ostracodes, and cystoids, with smaller amounts of tentaculitids and styliolinids, pelecypods, calcareous algae, calcareous worm tubes, gastropods, and sponge spicules. The other subtidal community is dominated greatly by massive stromatoporoids, but additionally contains abundant tabulate (*Favosites, Cystihalysites* and auloporoids) and rugose (*Lyrielasma* and *Tryplasma*) corals[5] and pelmatozoans (encrinal biosparite).

Samples of this facies are thin to medium-bedded. Numerous organisms of this habitat excreted the abundant fecal pellets. Intraclasts, denoting shallow conditions (Folk, 1959), generally are a small constituent of the rocks. Burrows, where present, are numerous and horizontal, often greatly disturbing bedding structures.

The extreme variability in sedimentary textures and structures represents several subenvironments in this third facies—ranging from a deeper-water, high-energy setting to a deeper-water, low-energy setting. Such subenvironments are reflected by: (1) the complete range of spar/micrite ratios—from biopelsparite (ratio greater than 2) to poorly washed biopelsparite (between 2 and ½) to biopelmicrite and biomicrite (less than ½); (2) the complete range of fossil abrasion—from well rounded to angular fossil fragments; and (3) the complete spectrum of sorting—from very good to very poor. These rocks were deposited on a subtidal sea floor, and the energy level varied from high to low. Water depth was only slightly greater than that of the shallow-water intraclastic rocks.

Summary Description of Environments

Using Irwin's (1965) model to aid in the interpretation of the Tonoloway, the three facies described above depict an onshore-to-offshore relationship (Fig. 3). The shallow subtidal facies is the rock record of the relatively high-energy belt where kinetic energy from waves acted upon the bottom sediment and tidal action was moderately strong. The deeper subtidal facies overlapped into the high-energy belt and extended seaward below wave base. Invertebrate life flourished within the waters of the high-energy belt (biopelsparite) but also thrived in the quiet, deeper waters of the adjacent low-energy zone (biopelmicrite and biomicrite), where marine currents were the only form of hydraulic energy acting on the bottom. Landward was the low-energy tidal flat, beyond any agitation of the high-energy belt. Little circulation and only minor tidal action marked the intertidal-supratidal facies. These same facies relations have been noted by Laporte (1971) in other Paleozoic epeiric seas of the Appalachians. Furthermore, the diversity levels and faunal compositions of the Tonoloway benthic communities closely parallel those described by Walker and Laporte (1970) from Ordovician and Devonian carbonates in New York.

Vertical distributions of both the microfacies and the benthic communities are plotted alongside the stratigraphic column of the Tonoloway at Pinto, Maryland (Fig. 4). The fit of petrographic and paleontologic data is very good, and both were used to arrive at environmental interpretations (last column). On the basis of this information, the Tonoloway Limestone has been divided informally into three members.

The lower member, 290 ft (88 m) thick, is the rock record of intertidal-supratidal deposition. It is characterized by micrite and pellets, thin bedding and laminations, stromatolites, mud cracks, gypsum molds, and the *Leperditia* and *Leperditia-Hormatoma* communities. A few beds of the shallow subtidal facies are present, formed by storms and (or) minor transgressions onto the mud flat. The middle member, 110 ft (33.5 m) thick, is extremely heterogeneous; the intertidal-supratidal, shallow subtidal, and deeper subtidal facies are all interbedded. This member is distinguished in the field

---

[5] Corals were identified by W. A. Oliver, U.S. Geological Survey, written communication.

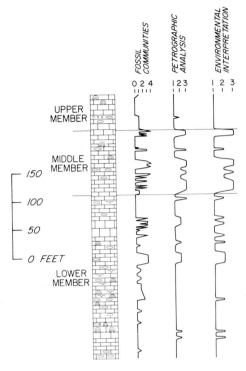

FIG. 4—Stratigraphic column of the Tonoloway Limestone at Pinto, Md. Symbols in lithologic column: circle, ooid; ellipse, intraclast; v, mud crack; convex curves, stromatolite. Fossil communities: 0, no fossils; 1, *Leperditia*; 2, *Leperditia-Hormatoma*; 3, *Leperditia-"Camarotoechia;"* 4, strophomenid brachiopod. Petrographic analysis: 1, micrite or pelmicrite; 2, intrasparite or intraclastic biosparite; 3, biopelsparite or biopelmicrite or biomicrite. Environmental interpretation: 1, intertidal-supratidal facies; 2, shallow subtidal facies; 3, deeper subtidal facies.

on the basis of the thicker bedding, intraclasts, "sparite" rocks, and high faunal diversity. Four communities are present but the stromatoporoid-coral assemblage did not become established at Pinto, Maryland. In general, a transgression was responsible for the more normal-marine conditions in the middle member, yet water depth fluctuated from zero to perhaps tens of feet. The upper member is approximately 200 ft (61 m) thick, but only 64 ft (19.5 m) is exposed. In the field and in thin section, this member is similar to the lower one. Moreover, only the *Leperditia* community is identified from these generally unfossiliferous beds. This member, though, is more argillaceous than the remainder of the formation, and we suspect that the covered interval is very shaly. The upper member also records intertidal-supratidal sedimentation—a regression following the transgression of the middle member.

Tonoloway Limestone Along the Outcrop Belt

Overall, the Tonoloway is identified in the outcrop on the basis of its very fine-grained, sparsely fossiliferous, argillaceous, and laminated lithology. These traits mark the upper and lower members as well as some beds of the middle member. The overlying Upper Silurian - Lower Devonian Keyser Limestone is clearly and easily separated from the Tonoloway. The Keyser, a "normal-marine" limestone, is very fossiliferous, thicker-bedded, often cherty, and generally has a knobby appearance at the base. On cross section A-A' (Fig. 5), along the tri-state outcrop belt (Fig. 1), fossils and cherty beds define the Keyser, the largest part of the Helderberg Group. (Ostracodes are not included in Fig. 5 because they are ubiquitous.) Succeeding cross sections illustrate that the Keyser retains these characteristics into the subsurface. The Wills Creek Formation, which underlies the Tonoloway, consists of interbedded calcareous shale and argillaceous limestone, and the upper contact is generally gradational with the Tonoloway. (For cross sections B-B' through E-E', the datum is the Mandata Shale, equivalent to the base of the upper cherty unit of the Helderberg. Because the Mandata is not well exposed in most of the outcrop area, the datum of cross section A-A' is the top of the Devonian Oriskany Sandstone.)

The three members of the Tonoloway can be observed readily along the outcrop belt (Fig. 5). On cross section A-A', the threefold division is visible, due to the presence of the fossiliferous middle member. Trace amounts of anhydrite (Pinto, Maryland) and dolomite (Lambert Gap, West Virginia, and Highland County, Virginia [Fig. 1]) identify the lower member, which contains some fossils other than ostracodes. The upper member is covered in large part at the three sections, presumably because of the large amounts of shale. Anhydrite is shown at Knapp Creek, West Virginia (Fig. 1); gypsum molds and halite-crystal impressions at Lambert Gap; and dolomite at both.

The same environments of deposition have been interpreted for the three members at all sections, spanning 120 mi (193 km). The lower and upper members are predom-

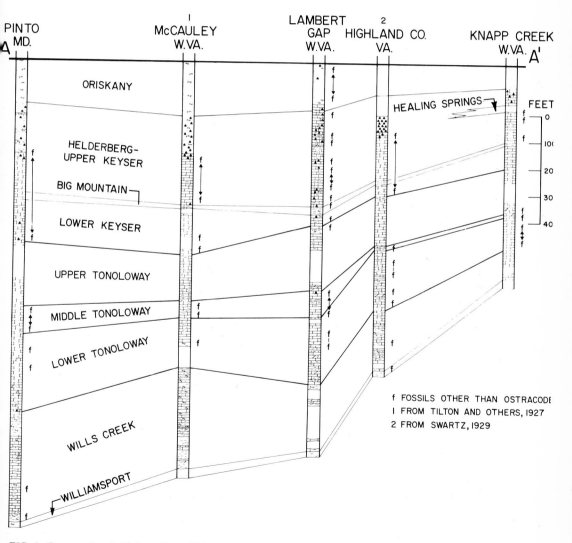

FIG. 5—Cross section A-A', from Pinto, Md., southward across outcrop belt. Location of cross section shown in Fig. 1.

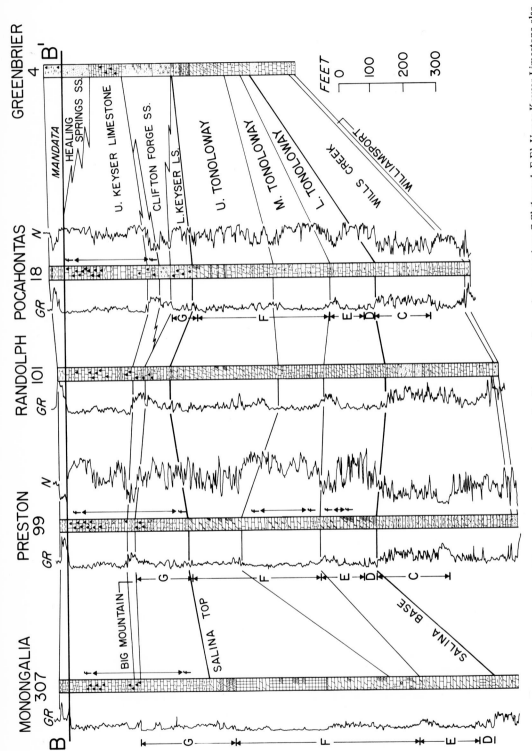

FIG. 6—Cross section B-B', beneath the eastern Appalachian plateau. On this cross section and cross sections C-C' through E-E', Upper Keyser Limestone also includes the overlying, thin New Creek and Corriganville Limestones; GR, gamma-ray log; N, neutron log; f, fossils (including ostracodes). Location of the cross

inantly of the intertidal-supratidal facies. The fossiliferous middle member with its variable lithology reflects small changes of sea level—as part of a widespread transgression—across a very low paleoslope.

Tonoloway Limestone of the Near Subsurface

Once the three-fold division of the Tonoloway Limestone, based on rock types and faunal content, was established for the surface sections, we attempted to extend it into the subsurface to determine any relationships between the Tonoloway and Salina formations. To accomplish this we used a two-step approach: first, to extend the members into the near subsurface, using lithologies and faunal contents noted in well-sample descriptions, and second, to extend them into the subsurface across West Virginia, using sample descriptions and gamma-ray log stratigraphy.

Both published (Martens, 1939 and 1945) and unpublished descriptions of samples from deep wells were used in extending the three members of the Tonoloway into the near subsurface along the eastern Plateau. At the same time, the relationships of these members to gamma-ray log, bulk-density log, and neutron log characters were noted. In tracing these correlations westward into Ohio, equal importance was given to gamma-ray log stratigraphy that has been used to divide the Salina Formation in the Michigan and Ohio basins (Clifford, 1973; Rickard, 1969; Ulteig, 1964). Rock types based on sample descriptions were plotted alongside the gamma-ray logs, and lithologic correlations and facies changes were interpreted. In all cases, the top of the Salina was determined on the basis of lithology; the boundary was placed at the top of the highest dolomite-anhydrite rock described from samples. For the most part, these samples were of good quality, being from older cable-tool or later rotary holes drilled with mud, not air[6].

From north to south on cross section B-B' (Fig. 6), the Tonoloway Limestone in the subsurface of the eastern Plateau thins significantly—from nearly 900 ft (274 m) in the Monongalia 307 well, to almost 700 ft (213 m) in the Randolph 101, to less than 250 ft (76 m) in the Greenbrier 4. Along this line the formation is composed of slightly different lithologies and hence represents environments of deposition different than those of the outcrop sections; however, the same vertical pattern (similar lower and upper members—in this case evaporites—and a fossiliferous middle limestone member) is present as in line A-A'.

In the Pocahontas 18 well (Fig. 6) the lower member is chiefly limestone, as along the Allegheny Front farther east, and the upper contact is placed at the top of a shaly limestone that also defines the top of the lower member in the Randolph 101. In other wells of line B-B' (Fig. 6), this member primarily is dolomite with only minor amounts of limestone. Dolomite and anhydrite have been logged in the two northernmost wells, and Salina D salt is present in the Monongalia 307 near the base of the interval (Fig. 6). Salinity of the paleoenvironment apparently increased toward the depocenter (Monongalia 307) of the sedimentary basin.

A distinctive limestone is in the middle of the Tonoloway. On the basis of sample descriptions, the middle member is predominantly a fossiliferous limestone (biosparite and biomicrite in the Pocahontas 18 and Preston 99 wells and probably in the intervening Randolph 101, from which good sample descriptions are not available. On the other hand, the middle member is a pelmicrite to the north, at least in part. One notable feature is that this member generally consists of interbedded limestone and dolomite, but also of limestone and dolomitic limestone (Randolph 101), and different types of limestone (Pocahontas 18). As in the surface sections, the middle member records several small-scale transgressions and regressions superimposed on the major transgression.

Southward from Randolph 101, anhydrite is restricted to the upper shalier member of the Tonoloway, which is also the most dolomitic. This dolomite-anhydrite sequence also extends northward, and in the Monongalia 307 there is almost 240 ft (73 m) of

---

[6]We are indebted to Mr. Glasko Rector of GeoLog Incorporated for many of these sample descriptions.

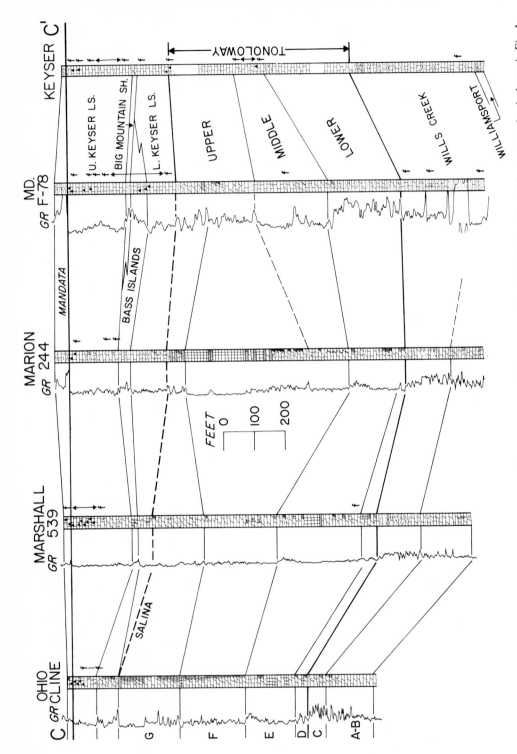

FIG. 7 –Cross section C-C', west to east, across the West Virginia salt basin and the eastern carbonate shelf. Location of the cross section is shown in Fig. 1.

Salina F salt. Excessive thickness of the F unit is attributed to salt flowage, a common occurrence farther north (Perry, 1975).

Both the Mandata Shale and the Big Mountain Shale Member of the Keyser Limestone provide key gamma-ray markers above the Tonoloway for correlation in the eastern Plateau (Fig. 6). Likewise, silty to sandy beds in the upper Wills Creek Formation below provide an equally good zone for correlation. However, drastic thinning southward through Greenbrier County (Fig. 6) renders correlations south of Pocahontas County suspect, and the middle member in Greenbrier 4 may be much thinner than shown. Correlations here are based on the assumption that anhydrite is restricted to the upper member (as in Pocahontas 18 and Randolph 101) and that the first silty-sandy section penetrated below the middle Tonoloway represents the upper Wills Creek.

In northeastern Greenbrier County, siltstone and sandstone occur near the top and base of the Tonoloway. Here and farther south, a southeastern to southern source for these clastic rocks is indicated. Indeed, a landmass existed in Virginia during Late Silurian through Early Devonian time (Dennison, 1970; Head, 1969) adjacent to the basin in West Virginia.

Salina Formation of the Subsurface

The threefold subdivision of the Tonoloway developed in the outcrop area (Fig. 5) and extended into the near subsurface along cross section B-B' (Fig. 6) can be continued farther into the Salina basin. Cross section C-C' (Fig. 7) trends from the eastern carbonate shelf through the West Virginia salt basin to the yoke between basins and terminates in the Ohio salt basin. The lower Tonoloway Limestone at the Keyser outcrop is replaced in western Maryland (F-78, Fig. 7) by calcareous dolomite and shale on the proposed shelf slope. Within the evaporite basin (Marion 244) the lower member becomes more dolomitic and anhydritic and is totally replaced by dolomite, anhydrite, and well-developed Salina D and E salts across the yoke (Marshall 539) into the Ohio basin (Ohio Cline). Thus, the lower member shows a progressive lateral facies change from limestone to dolomite to evaporites.

The middle member retains its highly calcareous, fossiliferous lithology into the center of the West Virginia salt basin (Marion 244, Fig. 7). Beyond this point anhydritic and saltiferous dolomite replace the middle member toward the Ohio basin. The contact between the middle and lower member, based on lithologic and faunal changes in surface exposures (line A-A', Fig. 5) and near-subsurface (line B-B', Fig. 6), is extended along the Salina E-F boundary of gamma-ray log stratigraphy (Fig. 7).

The upper member of the Tonoloway also illustrates lateral facies change from limestone to dolomite to evaporite from the outcrop to the center of the West Virginia salt basin. The Marion 244 well (Fig. 7) shows the tremendous thickness of Salina F salt in the basin—as much as 200 ft (61 m) in the upper member. The lower boundary of the upper member is chosen on lithologic change and is within the lower Salina F unit. The contact with the overlying Keyser Limestone is a lithologic change from dolomite and anhydrite to limestone. This lithologic boundary is progressively higher in the section from east to west based on gamma-ray log stratigraphy, placing the upper Salina contact at the top of the Salina G unit in Ohio. The evaporite sea retreated gradually into Ohio during late Cayugan time, while transgression of the Keyser sea flooded West Virginia; hence, the youngest anhydrites of the Salina Formation in Ohio are time equivalents of the oldest limestones of the Keyser Limestone in West Virginia.

This same basin-centered evaporite pattern is illustrated in cross section D-D' (Fig. 8) which also trends east-west, terminating on the western carbonate shelf (Wood 518 well). Limestone of the lower Tonoloway at the Lambert Gap outcrop is replaced progressively to the west by limestone and dolomite, then dolomite and anhydrite. No salts are developed in the lower member (as near the Ohio basin in line C-C') but the lower contact is marked by a clearly distinguished dolomite-anhydrite sequence at the base of the Salina D. Subsurface terminology in Ohio, based on gamma-ray log stra-

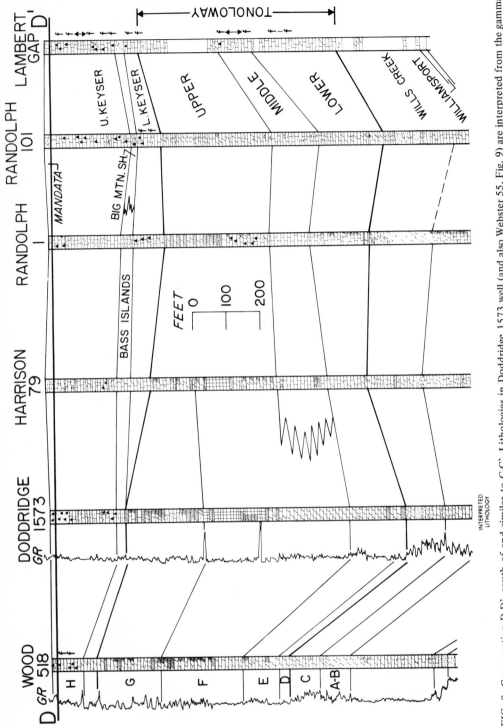

FIG. 8—Cross section D-D', south of and similar to C-C'. Lithologies in Doddridge 1573 well (and also Webster 55, Fig. 9) are interpreted from the gamma-ray character. Location of cross section is shown in Figure 1.

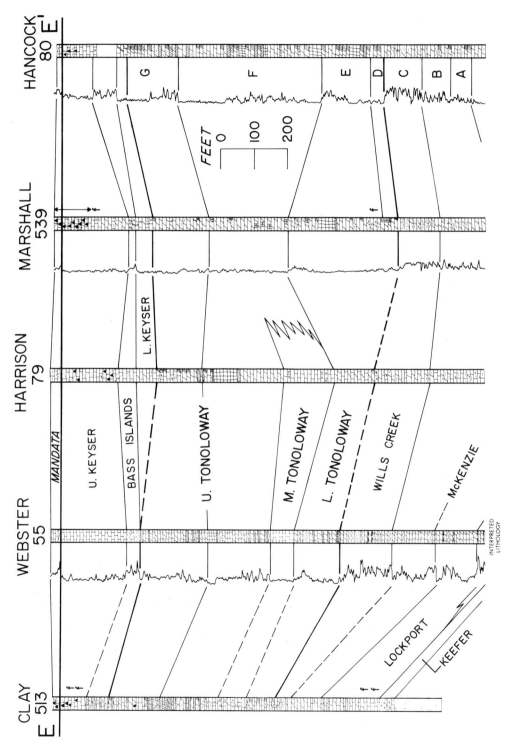

FIG. 9—Cross section E-E', across the southern carbonate shelf and into the West Virginia salt basin. Location of cross section is shown in Fig. 1.

tigraphy, is shown for the westernmost well, indicating a rough correlation of the lower Tonoloway with the Salina D plus E, the middle Tonoloway with the lower calcareous F, and the upper Tonoloway with the F and westward, all of the Salina G (Fig. 8). The middle Tonoloway Limestone member can be traced by lithology into the center of the salt basin (Harrison 79 well, Fig. 8). Farther west, dolomite and anhydrite are present in this interval. The upper member is a typical basin-centered evaporite, changing laterally from limestone to dolomite to dolomite-anhydrite and finally to salt. The top of the salt deposit is the approximate top of the Salina F. However, the uppermost anhydrites "climb" in the section, becoming younger to the west, and in Wood County, the top of the Salina is the top of the G unit.

Cross section E-E' (Fig. 9), extends from the Ohio basin (Hancock 80 well), across the yoke into the West Virginia salt basin, and terminates on the western shelf; the shelf is represented by oolitic limestones in the Clay County well. As on the other cross sections, correlations along line E-E' are based on lithologic changes, but correlations with gamma-ray markers mentioned previously also apply here. This line of section illustrates: (1) the western oolite shelf as a correlative of the middle Tonoloway Limestone member and (2) the different ages of the Ohio and West Virginia salt basins. Wells in and adjacent to the Ohio basin contain well-developed salts in the older D, E, and lower F units, whereas the main salt in West Virginia is uppermost F (Fig. 9). This progressive change from older salts farther west in Ohio to younger salts in West Virginia also is illustrated in both lines of section prepared by Clifford (1973).

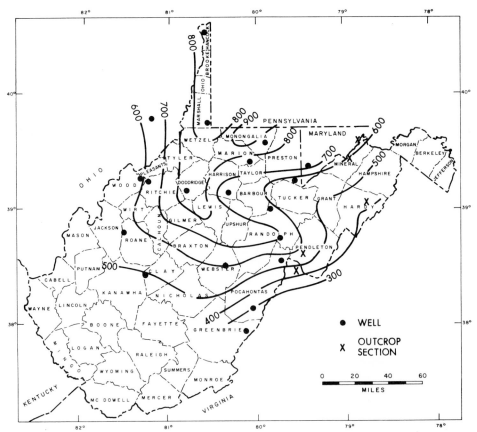

FIG. 10—Isopach map of the Tonoloway Limestone - Salina Formation. Contour interval is 100 ft (30.5 m).

Oolitic limestones and dolomites, often with shows of gas, have been penetrated in wells in Wood, Roane, Kanawha, Clay, and Fayette Counties (see Fig. 10 for county names). Although most of the gas encountered in the Salina is high in hydrogen sulfide, some shows of good gas have been tested, and one completion in the Salina Formation was attempted in Ritchie County (Fig. 10). The best possibilities for production of gas exist in oolitic shoal areas developed on the western shelf, and along the margin of the salt basin where carbonates and anhydrites locally pinch out updip against salt beds.

Paleogeography of Upper and Lower Members, Tonoloway Limestone

The five cross sections and the isopachous map of the Salina and Tonoloway (Fig. 10) outline the evaporite basin in northwestern West Virginia, surrounded to the east, south, and southwest by a carbonate shelf (Fig. 11). During deposition of the upper and lower members of the Tonoloway, salts precipitated in the basin (Salina Formation). Stromatolitic, pelletal, and ostracode-bearing micrites were deposited in intertidal-supratidal environments on the adjacent eastern carbonate shelf (Tonoloway Limestone). This mud flat was quite extensive along the Virginia landmass, being at least 120 mi (193 km) long. The slope was very gentle. Minor transgressions (shallow subtidal facies) onto the intertidal-supratidal mud flats are recorded in the upper and lower members of the Tonoloway. These fluctuations of sea level indicate increased water depth and reduced salinity, as shown by the moderate faunal assemblage.

The location, lithofacies, and form of the basin and shelf are best visualized in Figure 12. This illustration is a D-function-lithofacies map (Pelto, 1954) which delineates major regions based on the relative proportions of the four lithologies: limestone, dolomite, anhydrite, and halite. This map is of the upper member of the

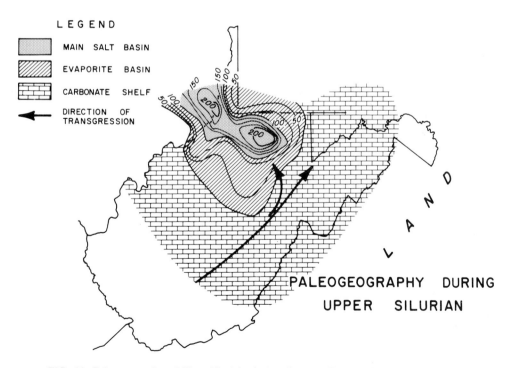

FIG. 11—Paleogeography of West Virginia during Cayugan Epoch, when the Tonoloway and Salina formations were being deposited. Open sea assumed to be southwest of the study area. Direction of transgression shown by arrow. Isopach lines indicate total Cayugan halite. Contour interval is 25 ft (7.6 m).

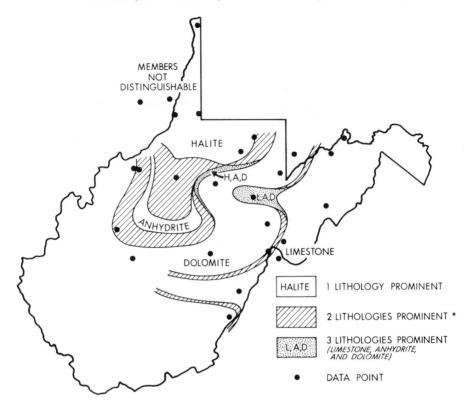

*Where two components are prominent in the stratigraphic interval, these must be the same as those lithologies of the adjoining sectors.

FIG. 12–D function-lithofacies map of the upper member of the Tonoloway-Salina interval. Map shows areal distribution of the four lithologic components (limestone, dolomite, anhydrite, halite). The three-fold division of the Tonoloway-Salina unit is not possible in Ohio and the northern panhandle of West Virginia.

Tonoloway-Salina unit, when the West Virginia evaporite basin was maximally developed. In the subsurface area of the eastern Plateau, two tongues of limestone, extended onto the shelf from the extreme east, are evident (one in the eastern panhandle of West Virginia, the other in the southeastern part of the state). The northern tongue coincides in position with a westward-projecting "thin" on the isopachous map (Fig. 10); the southern tongue follows the southern carbonate shelf (Fig. 11), and both parallel similar patterns in the underlying Williamsport Sandstone (projections of littoral sandstone into a carbonate basin [Patchen, 1973]). These tongues of limestone probably represent subtidal paleotopographic highs extending out toward the basin.

Easily noticeable on Figure 12 is the marked segregation of rock types in the concentric pattern characteristic of a closed basin (Schmalz, 1966). The basin was isolated from the open sea—restricted not by any sill, bar, or reef, but rather by a long fetch of very shallow water (Rickard, 1969). The climate must have been arid, though not necessarily hot, in order to maintain excessive evaporation. As the volume of brine was reduced and salinity increased, carbonates were deposited across the entire area (salina dolomite within the basin; Tonoloway tidal-flat limestone on the shelf), then anhydrite precipitated in the central region (within the basin and beneath the eastern plateau). Finally, halite formed in the deepest part of the basin. Hence, on a shelf-to-basin cross section or D-function - lithofacies map, the major lithologic component changes laterally from carbonate to sulfate to chloride.

The nature of the upper and lower members of the formations shows evidence of desiccating conditions, but the water mass never dried up completely. The several "cycles" of evaporite sedimentation seen in the Salina section (Figs. 6, 7 and 8) indicate that fresher water periodically flooded the basin. The open sea lay to the southwest (Head, 1969), and in all probability the brine was replenished occasionally with normal sea water from the south and southwest. The distribution of salts is somewhat asymmetrical—anhydrite to the south, halite to the north. This pattern may best be explained by proximity of the open sea to the southwest, with minor transgressions of fresher water from that direction. These "cycles" may be broadly correlative with the minor transgressions interpreted from outcrops of the Tonoloway, and such small-scale oscillations that affected both basin and shelf would point to the low paleoslope of the seafloor.

We envision, then, a broad, shallow basin as the site of accumulation of the evaporites—a sea desiccating with time. Water depth within the salt basin is a curious problem. Several facts indicate that the Salina salt basin of West Virginia was not deep: (1) Because halite appears to precipitate much faster than assumed rates of basinal subsidence (Schmalz, 1969), thickness of the halite deposit commonly is taken to be equal to or only slightly more than basin depth. The maximum thickness of any bed of halite in West Virginia is 110 ft (33.5 m) (and this value may be excessive due to ductile flow within anticlines). (2) The maximum rise of sea level during the major transgression represented by the middle member of the Tonoloway is thought to be quite small, on the order of several tens of feet. Yet this minor fluctuation was sufficient to alter profoundly the basinal deposits, which, we believe, also substantiates the argument for a shallow-water basin interior. (3) Likewise, Dellwig and Evans (1969) postulated that salts of the Salina of New York formed in a shallow, turbulent environment. The evidence for this conclusion is cross bedding, unconformities, shale balls, a shallow-water fauna, and absence of anhydrite-dolomite laminae.

The contour lines of Figure 11 depict the total thickness of halite, which ranges to a maximal 240 ft (73 m) in the Monongalia 307 well (Fig. 6, line B-B'). The greatest total thickness of the Cayugan deposits in the state (shown on the isopachous map, Fig. 10) is also in the Monongalia 307 well. In general, throughout the study area there is a high degree of similarity between distribution of halite and configuration of the basin.

Paleogeography of the Middle Member, Tonoloway Limestone

During deposition of the middle member of the Tonoloway, a major transgression deepened the sea over West Virginia. No evaporites were precipitated, but carbonates were deposited in the basin and on the surrounding shelf. Head (1969) postulated that the open sea lay southwest of West Virginia in Cayugan time; hence, the transgression probably came from that direction. The sea invaded the central Appalachian region along the eastern side of West Virginia and encroached upon the evaporite basin from southeast to northwest (Fig. 11). The basin was no longer restricted, and salinities were reduced drastically by the influx of normal sea water. The pattern of circulation was also modified and more mixing took place. Calcitic dolomite and limestone formed within the basin. The transgression, and the associated carbonate sedimentation, extended only into the West Virginia evaporite basin. In the northern panhandle of West Virginia and adjacent Ohio, anhydrite and halite were deposited continuously in the lower Salina F unit (Figs. 7 and 9), unaffected by freshing of the brine to the southeast. A yoke, postulated along the West Virginia - Ohio border, barred the Ohio region; constriction of contour lines in Marshall, Wetzel, Pleasants, and Wood Counties (Figs. 10, 11) hints at the location of this yoke. Northwest of the barrier, the threefold division of the Salina can not be recognized.

Subtidal limestones were deposited on the adjacent shelf. In the eastern outcrop belt, this is a fossiliferous limestone displaying high faunal diversity, with stromatoporoids and corals being the most noticeable. In the east there was no distinct shelf break, but instead, a very gradual slope. Sparse evidence of oolitic and fossiliferous limestones is present in several of the eastern wells (middle member in the Pocahontas 18, Harrison 79, and Monongalia 307 wells; upper member in the Greenbrier 4 well), as well as in the outcrop at Pinto, Maryland (upper member). However, no organic mounds or shelf-edge sands are present. To the south, on the other hand, was an oolitic shoal on the shelf edge (Figs. 9, 11). Deeper waters saturated with calcium carbonate flowed from the basin, over the shelf edge, and onto the platform. Along this edge, oolite bars developed within a high-energy, extremely shallow environment.

Although the most likely place for the occurrence of reefs was on the shelf or shelf edge during this interval of lowered salinity, none were penetrated by wells drilled through the section. Sea level fluctuated throughout the transgression, and the intermittent shallowing explains the lack of reef development in the Cayugan of West Virginia. The stromatoporoid-coral limestones of the outcrop belt are interbedded with intertidal-supratidal carbonates; these subaerial exposures repeatedly eliminated growth of the colonial organisms.

Summary

Figure 2 illustrates many of the stratigraphic relations summarized in this study. The Tonoloway Limestone is equivalent to the Salina Formation in the subsurface of West Virginia or the Salina D through G units of Ohio and Pennsylvania (Ulteig, 1964; Clifford, 1973). The lower member of the Tonoloway is correlated with the Salina D and E units; the middle member, with the lower Salina F. The upper Tonoloway member is equivalent to the upper Salina F of the eastern subsurface, and to the upper F and G units in the far west. The top of the Salina Formation rises in the section with respect to units in Ohio that are identified on the basis of gamma-ray character.

Evaporite-basin and enclosing carbonate-shelf environments of deposition are represented by the Salina and Tonoloway, respectively. Within the Salina Formation, halite is concentrated in the northern half of the basin, and anhydrite in the south. Several subenvironments of the carbonate shelf are recognized: intertidal-supratidal, subtidal topographic high, shallow subtidal, deeper subtidal, and oolitic environments. The overall distribution of lithofacies shows the salts to have been basin-centered, the characteristic pattern of a closed basin, restricted by a low paleoslope and the resulting long stretch of very shallow water. As the sea retreated into the shallow basin center, carbonates were deposited, followed by anhydrite and halite. However, periodic flooding of the basin came from the open sea to the southwest. A major transgression, identified in the middle member, temporarily upset this evaporite precipitation.

The depocenter of the evaporite basin shifted throughout the time of deposition. Salts of the Salina D and E are best developed in Ohio, and are present in West Virginia only in the westernmost counties. However, salts of the upper F unit are thickest to the east in northern West Virginia. Rickard (1969) demonstrated this same trend in the Salina of New York and Pennsylvania. Apparently, the entire axis of the Appalachian basin migrated to the east through the Cayugan Epoch.

Prospects for gas reservoirs include: (1) the oolite shoal on the carbonate shelf south of the evaporite basin, (2) local updip pinchouts of dolomite against salt beds within the basin, and (3) the stromatoporoid-coral facies of the middle member of the Tonoloway (which is well developed in Grant and Pendleton Counties) where it extends into the near subsurface, in Tucker, Randolph, and Pocahontas Counties (Fig. 10).

References Cited

Aitken, J. D., 1967, Classification and environmental significance of cryptalgal limestones and dolomites with illustrations from the Cambrian and Ordovician of southwestern Alberta: Jour. Sed. Petrology, v. 37, p. 1163-1178.

Alling, H. L., and L. I. Briggs, 1961, Stratigraphy of Upper Silurian Cayugan evaporites: AAPG Bull., v. 45, p. 515-547.

Cate, A. S., 1961, Stratigraphic studies of the Silurian rocks of Pennsylvania, Part 1, Stratigraphic cross sections of Lower Devonian and Silurian rocks in western Pennsylvania and adjacent areas: Pennsylvania Geol. Survey, Spec. Bull. 10, 3 p.

—— 1965, Stratigraphic studies of the Silurian rocks of Pennsylvania, Part 2, Subsurface maps of the Silurian rocks of western Pennsylvania and adjacent areas: Pennsylvania Geol. Survey, Spec. Bull. 11, 8 p.

Clifford, M. J., 1973, Silurian rock salt of Ohio: Ohio Geol. Survey, Rept. Inv. No. 90, 42 p.

Dellwig, L. F., and R. Evans, 1969, Depositional processes in Salina salt of Michigan, Ohio, and New York: AAPG Bull., v. 53, p. 949-956.

Dennison, J. M., 1970, Silurian stratigraphy and sedimentary tectonics of southern West Virginia and adjacent Virginia, in Silurian stratigraphy, Central Appalachian basin: Appalachian Geol. Soc., Charleston, West Virginia, p. 2-33.

Fergusson, W. B., and B. A. Prather, 1968, Salt deposits in the Salina Group in Pennsylvania: Pennsylvania Geol. Survey, Bull. M58, 41 p.

Folk, R. L., 1959, Practical petrographic classification of limestones: AAPG Bull., v. 43, p. 1-38.

Head, J. W., III, 1969, The Keyser Limestone at New Creek, West Virginia: an illustration of Appalachian Early Devonian depositional basin evolution, in A. C. Donaldson, ed., Some Appalachian coals and carbonates: models of ancient shallow-water deposition: West Virginia Geol. Survey, p. 323-355.

Irwin, M. L., 1965, General theory of epeiric clear water sedimentation: AAPG Bull., v. 49, p. 445-459.

Kinsman, D. J. J., 1969, Modes of formation, sedimentary associations, and diagnostic features of shallow-water and supratidal evaporites: AAPG Bull., v. 53, p. 830-840.

Laporte, L. F., 1971, Paleozoic carbonate facies of the central Appalachian shelf: Jour. Sed. Petrology, v. 41, p. 724-740.

Logan, B. W., R. Rezak, and R. N. Ginsburg, 1964, Classification and environmental significance of algal stromatolites: Jour. Geol., v. 72, p. 68-83.

Martens, J. H. C., 1939, Petrography and correlation of deep-well sections in West Virginia and adjacent states: West Virginia Geol. Survey, vol. XI, 255 p.

—— 1943, Rock salt deposits of West Virginia: West Virginia Geol. Survey, Bull. No. 7, 67 p.

—— 1945, Well-sample records: West Virginia Geol. Survey, Vol. XVII, 889 p.

Patchen, D. G., 1973, Stratigraphy and petrology of the Upper Silurian Williamsport Sandstone, West Virginia: West Virginia Acad. Sci., v. 45, p. 250-265.

Pelto, C. R., 1954, Mapping of multicomponent systems: Jour. Geol., v. 62, p. 501-511.

Perry, W. J., Jr., 1975, Thickness and extent of Silurian rocks, in W. de Witt, Jr., W. J. Perry, Jr., and L. G. Wallace, Oil and gas data from Devonian and Silurian rocks in the Appalachian basin: U.S. Geol. Survey, Misc. Invest. Series, Map I-917B.

Rhoads, D. C., 1967, Biogenic reworking of intertidal and subtidal sediments in Barnstable Harbor and Buzzards Bay, Massachusetts: Jour. Geol., v. 75, p. 461-476.

Rickard, L. V., 1969, Stratigraphy of the Upper Silurian Salina Group New York, Pennsylvania, Ohio, Ontario: New York State Museum and Sci. Serv., Map and Chart Series No. 12, 57 p.

Schmalz, R. F., 1966, Environments of marine evaporite deposition: Mineral Industries, v. 35, p. 1-7.

—— 1969, Deep-water evaporite deposition: a genetic model: AAPG Bull., v. 53, p. 798-823.

Swartz, C. K., et al, 1923, Silurian: Maryland Geol. Survey, 794 p.

Swartz, F. M., 1929, The Helderberg Group of parts of West Virginia and Virginia: U.S. Geol. Survey, Prof. Paper 1580, p. 27-75.

Tilton, J. L., et al, 1927, Hampshire and Hardy Counties: West Virginia Geol. Survey, 624 p.

Ulteig, J. R., 1964, Upper Niagaran and Cayugan stratigraphy of northwestern Ohio and adjacent areas: Ohio Geol. Survey, Rept. of Invest. No. 51, 48 p.

Walker, K. R., and L. F. Laporte, 1970, Congruent fossil communities from Ordovician and Devonian carbonates of New York: Jour. Paleontology, v. 44, p. 928-944.

Woodward, H. P., 1941, Silurian System of West Virginia: West Virginia Geol. Survey, Vol. XIV, 326 p.

Carbonate-Anhydrite Facies Relationships, Otto Fiord Formation (Mississippian-Pennsylvanian), Canadian Arctic Archipelago[1]

GRAHAM R. DAVIES[2]

Abstract The Otto Fiord Formation is a major evaporite deposit of Late Mississippian to Middle Pennsylvanian age that occupies an axial position in the Sverdrup basin of the Canadian Arctic Archipelago. Where it is exposed in normal stratigraphic succession on northwestern Ellesmere Island, the formation is composed of 350 to 600 m of rhythmically-alternating limestone and anhydrite, with some sandstones interbedded near the top. The formation is underlain by continental to marginal-marine red beds, and overlain by carbonates and shales of deep-water aspect. Laterally, the Otto Fiord evaporites and limestones are coeval with marine shelf carbonates that flank margins of the evaporite-carbonate "basin." In the central and southern Sverdrup basin, a facies of salt and anhydrite is buried by thick Mesozoic rocks and revealed at the surface only in large diapiric structures.

At several localities on northwestern Ellesmere Island, limestone mounds are enclosed within the Otto Fiord Formation. At van Hauen Pass, mounds as thick as 30 m and as long as 350 m are built on erosional remnants of anhydrite. The mounds have a crinoid-rich marine limestone base, a main beresellid algal-buildup facies of hypersaline aspect, and several marine limestone capping beds. The marine limestone phases of the mounds have thin, marine limestone equivalents in the intermound setting, separated from each other by units of anhydrite.

Reconstruction of depositional events reveals that the mounds are the composite record of at least three transgressive-regressive cycles, in which limestones were deposited during the marine phase and the algal facies and anhydrite units were formed under hypersaline conditions imposed by evaporitic drawdown.

In the Otto Fiord Formation of northwestern Ellesmere Island, in a facies change from evaporites to basin-margin reef and shelf carbonates, anhydrite units are interbedded with limestones at the feet of steeply-dipping tongues of shelf-foreslope carbonate rock. Depositional relief on these carbonate units increases upward through the evaporite section from a few meters to more than 400 m. Anhydrite interbedded with the limestones apparently formed subaqueously under conditions of depressed hypersaline sea level, in water probably shallower than depths suggested by depositional relief on the carbonates, but far from a sabkhalike environment.

The complex facies within the Otto Fiord Formation may serve as guides elsewhere in modeling evaporite-reef relations and major carbonate-evaporite facies transitions, and in studies of the subsurface related to exploration for hydrocarbons.

Introduction

The high-relief fiord walls and Arctic climate of northwestern Ellesmere Island (Fig. 1) combine to provide many spectacular exposures of upper Paleozoic rocks in which facies relations are revealed clearly. These conditions are particularly advantageous for study of sulfate evaporite rocks, because anhydrite commonly is preserved in exposures with little or no hydration to gypsum. Partly because of this excellent preservation, Davies and Nassichuk (1972, 1975) and Wardlaw and Christie (1975) were able to document evidence of a submarine depositional origin of evaporite strata in the Otto Fiord Formation (Upper Mississippian to Middle Pennsylvanian).

In this paper, emphasis is placed on documentation of the lithologic composition and depositional relations of facies-equivalent anhydrite and carbonate units of the Otto Fiord Formation on northwestern Ellesmere Island. Two major facies types are described—algal mounds (reefs) enclosed by anhydrite, and a shelf or basin-margin transition from carbonate to anhydrite.

[1] Manuscript received November 16, 1976; accepted March 7, 1977.

[2] Formerly, Institute of Sedimentary and Petroleum Geology, Geological Survey of Canada, Calgary, Alberta. Currently, Geological consultant, 4216 Varmoor Road, NW, Calgary, Alberta, Canada T3A 0B3.

Field work that included study of the Otto Fiord Formation was conducted in 1971, 1972, and 1974, in cooperation with W.W. Nassichuk, Geological Survey of Canada. The stratigraphic framework for this and other sedimentologic studies by the author in the Arctic was provided by the pioneering work of Ray Thorsteinsson, Geological Survey of Canada. I thank R. A. Meneley and The Canadian Society of Petroleum Geologists for permission to republish Figure 3.

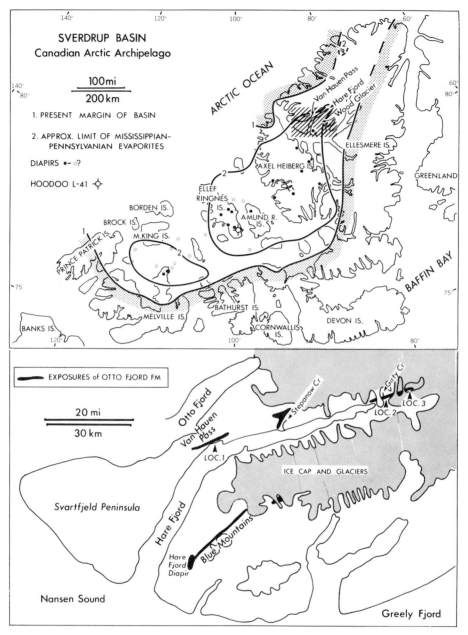

FIG. 1—Locality maps of the Arctic Archipelago and northwestern Ellesmere Island, showing subsurface distribution of evaporites in the Otto Fiord Formation, as defined by diapirs (after Meneley et al, 1975), and local outcrop belts of evaporites on northwestern Ellesmere Island (lower map). Sites on Ellesmere Island described in this paper are localities 1, 2 and 3, lower map. The Hoodoo L-41 well on Ellef Ringnes Island is the only well that has penetrated a halite facies of the Otto Fiord Formation.

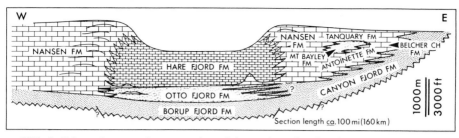

FIG. 2—Schematic restored cross section of the northern Sverdrup basin, northwestern Ellesmere Island, including rocks of Late Mississippian to Early Permian age. The section emphasizes the axial distribution of the Otto Fiord evaporites, and the major shelf-to-trough facies transition above the evaporites.

Stratigraphy and Regional Setting, Otto Fiord Formation

The Otto Fiord Formation is part of the sedimentary fill of the Sverdrup depositional basin (Thorsteinsson, 1974), an elongate depression that formed in Mississippian time by crustal thinning along the axis of the precursor Franklinian "geosyncline." The basin is filled by sedimentary rocks that range from Mississippian to Cretaceous, with extensive intrusive and minor extrusive igneous rocks as additional components. In general, the Mississippian to Permian succession is an association of red beds, evaporites, carbonates, and cherty argillites, whereas the Triassic to Cretaceous succession is shale and sandstone.

Thorsteinsson (1974) provided a stratigraphic framework for upper Paleozoic rocks of northwestern Ellesmere Island and other areas of the northern Arctic Archipelago. Exposures of the Otto Fiord Formation in normal stratigraphic succession are confined to two outcrop belts on northwestern Ellesmere Island (Fig. 1). Where the upper Paleozoic succession is complete, the formation is underlain by about 400 m of Upper Mississippian conglomeratic red beds of the Borup Fiord Formation (Fig. 2). The contact between the two formations commonly is gradational, with red sandstones of the Borup Fiord interbedded with carbonates of the basal Otto Fiord. Discrete lenses of oolitic and bioclastic limestone are within the uppermost red beds of the Borup Fiord. In some sections, however, there is evidence of local unconformity between the two formations, or of faulting in red beds of the Borup Fiord, before deposition of evaporites. These variations are consistent with the structural and environmental evolution of the Sverdrup depression from dominantly nonmarine to marine, with active tectonism and subsidence along fault zones.

The Otto Fiord Formation is overlain conformably by 500 m or more of Middle Pennsylvanian to Lower Permian argillaceous and cherty carbonates and shales of the Hare Fiord Formation. These rocks were deposited in deeper waters of the Sverdrup trough; they contain carbonate turbidites, debris sheets and other displacement phenomena (Davies, in press). The Hare Fiord Formation grades laterally into coeval shallow-water shelf carbonates of the Nansen Formation (Fig. 2) that are up to 2,000 m thick (Thorsteinsson, 1974; Davies, 1975a).

The Otto Fiord Formation on northwestern Ellesmere Island is maximally about 500 m thick. It is composed of rhythmically alternating limestone and anhydrite units that are interbedded with mudstone, shale and sandstone in the upper part of the formation. At several localities, limestone mounds or reefs as thick as 30 m are enclosed in anhydrite of the Otto Fiord. Limestone units characteristically are skeletal wackestones with a varied marine biota and only traces of dolomite; commonly they are 1 to 12 m thick, and thickness and biotic variety of units increases up-section. Anhydrite strata may be bedded or laminated, or have a nodular-mosaic fabric in which bedding is poorly preserved; pseudomorphs after a primary crystalline gypsum fabric are preserved in some beds (Davies and Nassichuk, 1975, Fig. 6; Wardlaw and

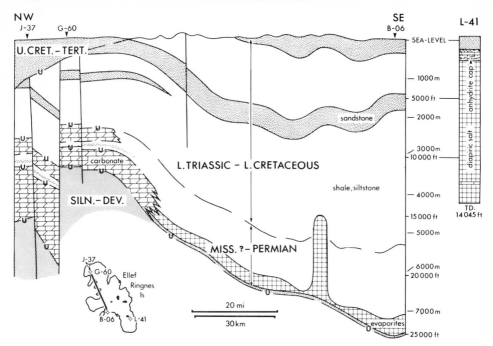

FIG. 3—Interpretive section across northwestern and western Ellef Ringnes Island, based on seismic and well data (modified from Meneley et al, 1975), with an appended lithologic summary of the Hoodoo L-41 well. The section emphasizes the faulted, uplifted northwestern rim of the Sverdrup basin, buildup of upper Paleozoic carbonate sediments on this uplift, and a halite facies of the Otto Fiord Formation. The Hoodoo L-41 well was drilled off the flank of the exposed Hoodoo diapir; it penetrated an anhydrite-carbonate cap and 3,700 m of salt.

Christie, 1975, Figs. 15-18). Anhydrite units commonly are 3 to 50 m thick, with some exceeding 60 m.

The Otto Fiord Formation ranges from Late Mississippian to late Middle Pennsylvanian. In North American stage terminology, the range is from late Chesteran to late Desmoinesian; in European terminology, from earliest Namurian to middle Moscovian (Nassichuk, 1975). Within this time bracket, however, are diachroneities and hiatuses between various sections and facies.

At a few localities along northern Hare Fiord (Fig. 1), evaporites of the Otto Fiord can be traced through a facies change into carbonate rocks of shelf-edge or basin-margin aspect. At Stepanow Creek (Fig. 1), these coeval carbonate rocks are the basal component of the thick Nansen shelf carbonate and appear to be of shallow-water depositional aspect. East of Girty Creek (Fig. 1), anhydrite and other (interbedded) rocks of the Otto Fiord Formation reveal a complex facies and structural relation with carbonates, which includes steeply dipping shelf-edge or reef-foreslope beds of the Nansen Formation.

Elsewhere in the Sverdrup basin, rocks of the Otto Fiord Formation are buried by thick Mesozoic and Tertiary sediments. However, presence of evaporites of the Otto Fiord below this cover is revealed by more than 110 diapirs and intrusive bodies of anhydrite exposed on five of the Arctic Islands. Fossils from limestone interbeds exposed within anhydrite at the surface in the diapirs confirm the correlation with the Otto Fiord on northwestern Ellesmere Island. The diapir belt and the Ellesmere exposures together define an area underlain by evaporites that is 800 km long and as wide as 240 km.

Although anhydrite with surficial gypsum is the only evaporite phase exposed at the surface in the diapirs, some geophysical data have suggested a halite core (see Thorsteinsson, 1974, p. 74-81, for a review of literature on diapirs of the Otto Fiord Formation). Direct proof of a halite core was provided by the Hoodoo L-41 exploratory well drilled in 1972 off the flank of Hoodoo diapir on Ellef Ringnes Island (Fig. 1); this well penetrated almost 3,700 m of halite before being abandoned (Davies, 1975b). Geometry and depth of burial of the Otto Fiord salt facies and diapirs have been illustrated by Meneley et al (1975, Figs. 4, 6). From one of these cross sections (Fig. 3) it is apparent that the northwestern rim of the Sverdrup basin was tectonically active and probably was a bathymetric high during part, if not all, of the early phases of subsidence of the Sverdrup trough. This rim thus formed the western to northwestern boundary of the Sverdrup depression during deposition of the Otto Fiord evaporites and during subsequent depositional events. The same style of fault-controlled structural high probably extended northward across northwestern Ellesmere Island, and confined the deposition of evaporites. If evaporites of the Otto Fiord Formation are linked to lithologically similar and age-equivalent evaporites on Spitzbergen and Greenland, they seem to define a regional trend of evaporites—albeit not necessarily continuous—that may have been deposited within a major late Paleozoic rift-like system.

Evaporites of the Otto Fiord within the Sverdrup basin probably accumulated in a number of separate or interconnected sub-basins. Meneley et al (1975, Fig. 2) suggested that salt deposition occurred in two sub-basins defined now by the bimodal areal distribution of major diapirs. On northwestern Ellesmere Island, the youngest evaporites in the two major outcrop belts, a little more than 30 km apart, differ in age from early Atokan (Hare Fiord area) to middle or late Desmoinesian (northern Blue Mountains area, Fig. 1) (W. W. Nassichuk, personal commun., 1976). This degree of diachroneity over a relatively short distance is explained most readily by continued deposition of evaporites in the northern Blue Mountain area behind a structural or reefal carbonate barrier, after onset of more normal marine conditions in the Hare Fiord area.

Depositional Environment of Anhydrite in the Otto Fiord Formation

Evaporites of the Otto Fiord Formation are of interest because of their exceptional exposure and preservation, and also for the insight they provide into depositional environments. Davies and Nassichuk (1972, 1975) and Wardlaw and Christie (1975) concluded that anhydrite strata of the Otto Fiord were deposited in a submarine or hypersaline subaqueous environment. This interpretation contrasts specifically with the currently popular sabkha or paleosabkha depositional model applied to explain many other ancient evaporite units. The evidence for interpretation of submarine deposition is discussed thoroughly by Davies and Nassichuk (1975), Wardlaw and Christie (1975), and Nassichuk and Davies (in preparation); therefore, it is discussed here only briefly and is summarized in tabular and schematic form (Table 1; Figs. 4, 5).

The fundamental aspect of this interpretation is that most if not all of the sulfate evaporites in the Otto Fiord Formation accumulated by precipitation, primary crystal growth, and early diagenetic emplacement of $CaSO_4$ beneath a hypersaline water mass. This water mass probably was stratified compositionally, but was not necessarily "deep." The hypersaline condition developed primarily through a combination of tectonic and climatic controls, and resulted in lowering of the saline water level by "evaporative drawdown" (Maiklem, 1971). Evaporites thus accumulated in shallower water than the interbedded marine limestones. However, even if a particular evaporitic pulse culminated in total desiccation, it is evident that the great mass of evaporitic minerals must have precipitated from the increasingly concentrated brine as a subaqueous product, rather than at an emergent, sabkhalike surface. A necessary corollary of this interpretation is that coeval carbonate and other rock types peripheral to the evaporite depocenter were exposed periodically during the evaporitic maxima.

Although interpretation of depositional environments of the Otto Fiord Formation has been based on distribution and lithologic characteristics of anhydrite units and associated carbonates, the same interpretation is extended in principle to facies-equivalent salt in the subsurface of the central and southern Sverdrup basin.

Sandstone interbedded with anhydrite and limestone of the Otto Fiord Formation on Ellesmere Island characteristically is composed of calcite-cemented quartz sand with marine bioclasts (mainly foraminifers) and sparse wood fragments. Although at least some of the bioclasts may be reworked, these sandstones and associated shales and mudstones probably also were deposited in a marine (hypersaline) environment, albeit at the shallowest phase of the depositional cycle (see Davies and Nassichuk, 1975, Figs. 5B, 7).

Limestone Mounds in the Otto Fiord Formation

At several localities along Hare Fiord (Fig. 1), limestone mounds or reefs of Early Pennsylvanian age are exposed in the Otto Fiord Formation. The more spectacular of these mounds are in the type area near van Hauen Pass (Fig. 1). The general section at van Hauen Pass consists of 400 m of anhydrite and limestone of the Otto Fiord

TABLE 1. CHARACTERISTICS OF OTTO FIORD MARINE CARBONATE-ANHYDRITE CYCLES AND IDEALIZED SABKHA CYCLE

Otto Fiord Cycle (Submarine Environment)	Sabkha Cycle (Subaerial Environment)
1. Basin axis distribution	1. Basin margin (shelf) distribution
2. Shallow marine carbonate rocks "landward" of evaporites	2. Terrigenous-continental rocks "landward" of evaporites
3. Major salt facies in basin center(?)	3. Marine (carbonate) facies in basin center
4. Thick cycles common (10 to 50+ m)	4. Thin cycles common (1 to 3 m)
5. High depositional/topographic relief on some interbedded carbonate rocks (30 to 400+ m)	5. Low depositional/topographic relief for sabkha complex
6. Marine biota in carbonate units, continuing up to anhydrite contact; biotic variety increases upward through unit	6. Marine biota in basal (subtidal) carbonate units, decreasing in variety and volume into intertidal subfacies
7. Dolomite not common; late diagenetic where present	7. Dolomite common; early diagenetic
8. Carbonate-anhydrite transition marked by thin, parallel laminae of both rock types; stromatolites and desiccation structures absent	8. Intertidal transition marked by algal laminae (irregular), stromatolites, desiccation structures
9. Discrete nodules and enterolithic layers of anhydrite absent at carbonate to anhydrite transition	9. Nodules and enterolithic layers of anhydrite common at transition and in supratidal facies
10. Regular, parallel bedding 10 to 15 cm thick common in otherwise massive anhydrite units	10. Regular thick bedding not found in sabkha deposits
11. Fabrics in anhydrite beds pseudomorphic after large, vertical to inclined gypsum crystals	11. Beds of subvertical gypsum crystals not found in sabkha deposits
12. Some thin-bedded and parallel-laminated anhydrite	12. Thin-bedded and parallel-laminated anhydrite uncommon or absent
13. Nodular mosaic fabric common in thick anhydrite cycles; considered to be secondary (diagenetic) fabric	13. Coalesced nodular fabric present; primary to very early diagenetic

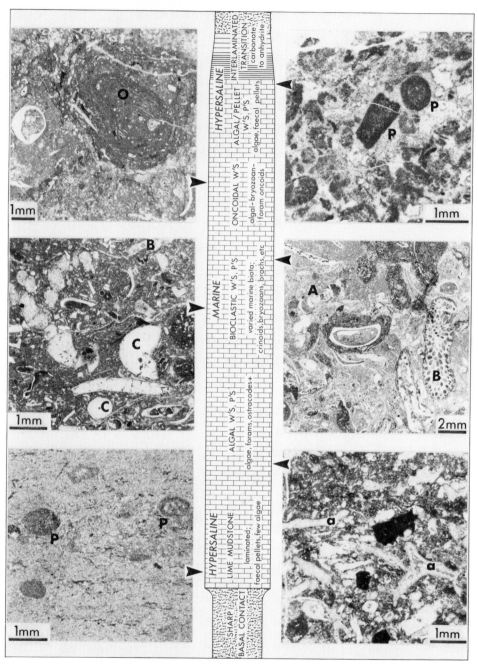

FIG. 4–Lithologic succession in limestone unit of a typical Otto Fiord carbonate-anhydrite cycle, northwestern Ellesmere Island. The complete succession is interpreted to record a major marine transgressive event that began with accumulation of abiotic carbonate sediment under hypersaline (mixed source?) conditions, and graded through euryhaline biotic conditions into a more normal marine environment that supported a varied biota. Uppermost in the section are rocks indicating increasingly hypersaline conditions, and the onset of deposition of evaporites, probably accompanied by evaporative drawdown of water level. Symbols: W'S—wackestone; P'S—packstone; C—crinoid; A—ammonoid; B—bryozoan; a—algae; P—fecal? pellet; O—oncoid.

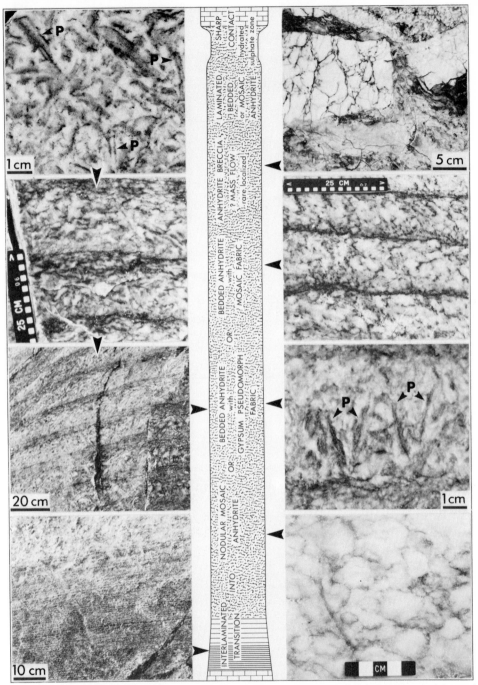

FIG. 5—Lithologic succession in anhydrite unit of a typical Otto Fiord carbonate-anhydrite cycle, northwestern Ellesmere Island. Transition from a carbonate unit (Fig. 4) into anhydrite commonly is gradational and composed of interlaminated limestone and anhydrite. Many anhydrite beds are composed almost entirely of nodular-mosaic fabrics, but others show a variety of primary bedding structures and laminae. Fabrics pseudomorphic after primary or early diagenetic gypsum crystals (P) are in some beds. Anhydrite breccias, possibly of mass-flow or slump origin, are in some units, but are very rare. Bounding carbonate strata were aquifers during erosional unloading of these rocks, resulting in development of gypsum immediately below and commonly above the limestone units.

FIG. 6—Exposure of Otto Fiord Formation at the type area near van Hauen Pass, Ellesmere Island. Note rhythmic interbedding or cyclicity of dark-colored limestone and light-colored anhydrite beds, and three algal mounds (arrows) exposed at mid-section. Load-deformation of units below mound at far left is defined clearly. Dark-colored rocks above the Otto Fiord evaporites are shales and cherty carbonates of the Hare Fiord Formation. Discontinuous sandstone units (S) are between the mounds.

Formation overlain by shales and cherty carbonates of the Hare Fiord Formation. Lower units of the Otto Fiord are faulted out and missing. The Otto Fiord section is a rhythmic succession of interbedded limestone and anhydrite beds that generally increase in thickness up-section (Fig. 6).

About 200 m above the fault-bounded base of the Otto Fiord section, four limestone mounds are exposed within a limestone marker bed (Fig. 6). The largest mound is about 30 m thick and 350 m long in the plane of exposure. The mounds have flat bases and irregular upper profiles (Fig. 7). Variations in weathering character of internal components reveal that the mounds are composite buildups with three or more stacked and thickened limestone units forming each mound (Fig. 7). Apparent depositional dips at flanks of the mounds and of internal components are as steep as 25° to 30°. In the intermound area, each of the main lenticular limestone units of the mound is represented by thinner, planar beds of limestone separated by units of anhydrite (Fig. 7).

The mounds are built on elevated blocks or platforms of anhydrite that have relief of about 14 m above the surface of anhydrite in the intermound areas (Fig. 7). These anhydrite platforms are interpreted to be erosional remnants formed during a period of subaerial exposure. Geometry of the irregular surface is defined best by using the thick limestone unit that lies about 30 m below the mounds (Fig. 7) as a horizontal datum. Two thin limestone marker beds that are overlain by about 15 m of anhydrite below the centers of the mounds appear to be truncated at the flanks of the anhydrite platform (Fig. 7). This apparent truncation is the main criterion for interpretation of an erosional origin of the anhydrite platforms. Preservation of regular, parallel bedding

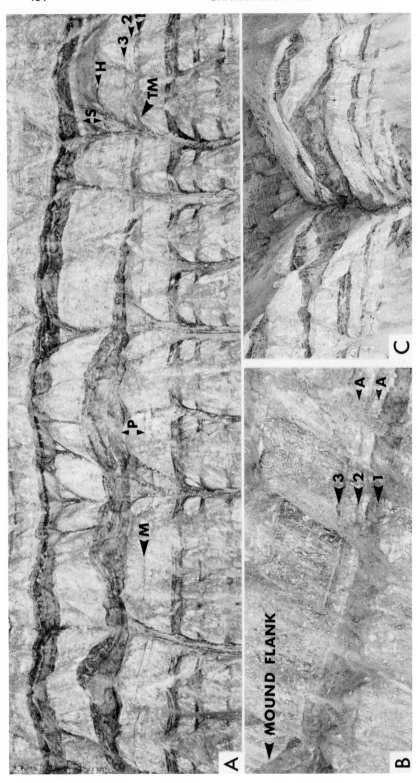

FIG. 7.—Lower Pennsylvanian algal mounds in the Otto Fiord Formation near van Hauen Pass. **A.** General view of mound shown at far left of Figure 6. Mound is about 30 m thick. Note strong cyclicity and general increase in thickness of successive limestone and anhydrite units. At least the upper of the two thin limestone marker beds (M) below the mound is truncated below flanks of the mound (TM). The mound is built on an apparent anhydrite "platform" about 14 m thick (P). Internal bedding, some steeply dipping, is evident within the mound. One thick and two thin limestone units (1-3) with interbedded anhydrite form the lateral equivalent of the mound in the flank setting. A discontinuous hematite zone (H) and a sandstone unit (S) are enclosed in the thick anhydrite unit in the intermound area. **B.** Detail of mound flank showing the basal limestone unit (1), two thinner limestone beds (2, 3) that lap onto flanks of the mound, and anhydrite beds (A) between limestones. **C.** Algal mound shown at far right (east) of Figure 6.

in at least part of the anhydrite of the platform unit supports this interpretation, and specifically argues against the alternative concept of mechanical construction of the platforms as banks of detrital sulfate on a more-or-less horizontal substrate.

Accumulation of carbonate sediment after erosion of the underlying anhydrite was controlled to a large degree by geometry of the eroded surface. Mound building was localized on the platformlike surface of the erosional remnants of anhydrite. Thickest accumulations appear to overlie slight topographic highs on the platform. Abrupt thinning of limestone flank beds and the nearest off-mound anhydrite interbeds is evident above flanks of the anhydrite platform (Fig. 7).

In the off-mound or intermound setting, the equivalent section is composed of three limestone beds that are lateral extensions of thicker accretionary units in the mound (Fig. 7). The beds are separated by anhydrite units that pinch out between the limestone beds at the mound flanks, with the upper anhydrite units onlapping further and higher onto the flanks.

The mounds are buried by anhydrite that is as thick as about 50 m in off-mound settings. This anhydrite is overlain by a prominent unit of shale and limestone 20 m or more thick. This limestone unit is as much as 5 m thicker where it overlies the limestone mounds (Wardlaw and Christie, 1975, p. 160).

A discontinuous tabular sandstone unit is in the section about 30 m above the uppermost limestone bed in the off-mound area (Fig. 7). This sandstone is about 18 m thick, and is composed mainly of cross-bedded and ripple-laminated calcite-cemented quartz sand. It contains carbonized plant fragments and marine skeletal grains. Green shale and siltstone beds are common in the lower part of the unit.

Close to the flanks of the mounds, the sandstone unit thins abruptly and is represented over the flanks only by a thin, green argillaceous marker bed that can be traced from the top of the sandstone unit some distance through the anhydrite above the flanks and tops of the mounds (Fig. 7). Thinning of the sandstone beds near the mound flanks appears to proceed from the base of the unit upward, consistent with the geometry expected for a sedimentary infill between mounds. The irregular surface defined by the sandstone bodies and the thin green marker beds over the mounds is believed to correspond with depositional topography on the underlying anhydrite. Significantly, this would require deposition of anhydrite in water at least several tens of meters deep, and thus is consistent with the submarine depositional model discussed earlier.

The discontinuous sandstone bodies are underlain by a very prominent but also discontinuous bed of red hematite and hematitic anhydrite (Fig. 7). This bed is 2 to 3 m thick, and is separated from the base of the overlying sandstone by about 7 m of crumbly-weathering anhydrite. The hematite bed thins and pinches out off the flanks of the limestone mounds (Fig. 7), thus conforming to the same distributional pattern as the sandstone units. This bed is interpreted to record a period of accumulation and preservation of iron oxides, probably originally in hydrous mineral form, on the deeper floor of the intermound depressions.

The discontinuous hematite and sandstone bodies are overlain by several meters of anhydrite that are continuous across tops of the limestone mounds. Thus the complex series of depositional events in the intermound or off-mound setting terminated in a return to accumulation of sulfate evaporite in a continuous blanket with little depositional relief. The entire succession is overlain by a limestone unit 20 m thick.

Internal Lithologic Succession

Limestone mounds in the Otto Fiord Formation at van Hauen Pass are characterized internally by a vertical lithologic sequence from crinoid-rich skeletal wackestones at the base, through the main algal-mound phases to, caps of skeletal wackestone with a varied marine biota. In more detail, the sequence from the base upward (Fig. 8) is as follows:

1. Sharp contact with underlying anhydrite (or granular gypsum in hydrated zone within 1 m of contact), with some irregularities apparently caused by local filling of sediment in depressions on the erosion surface.

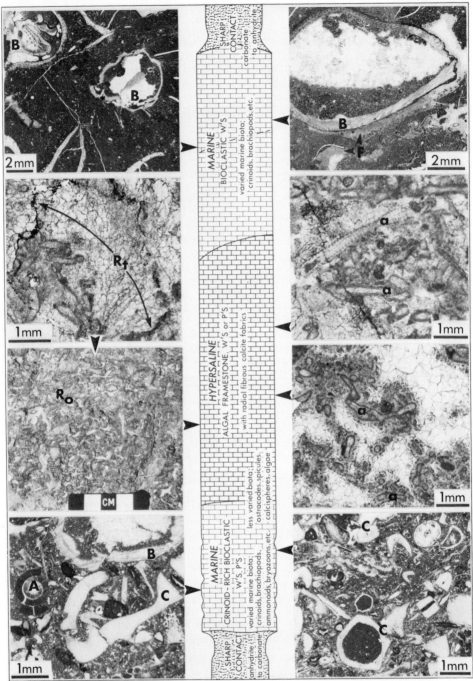

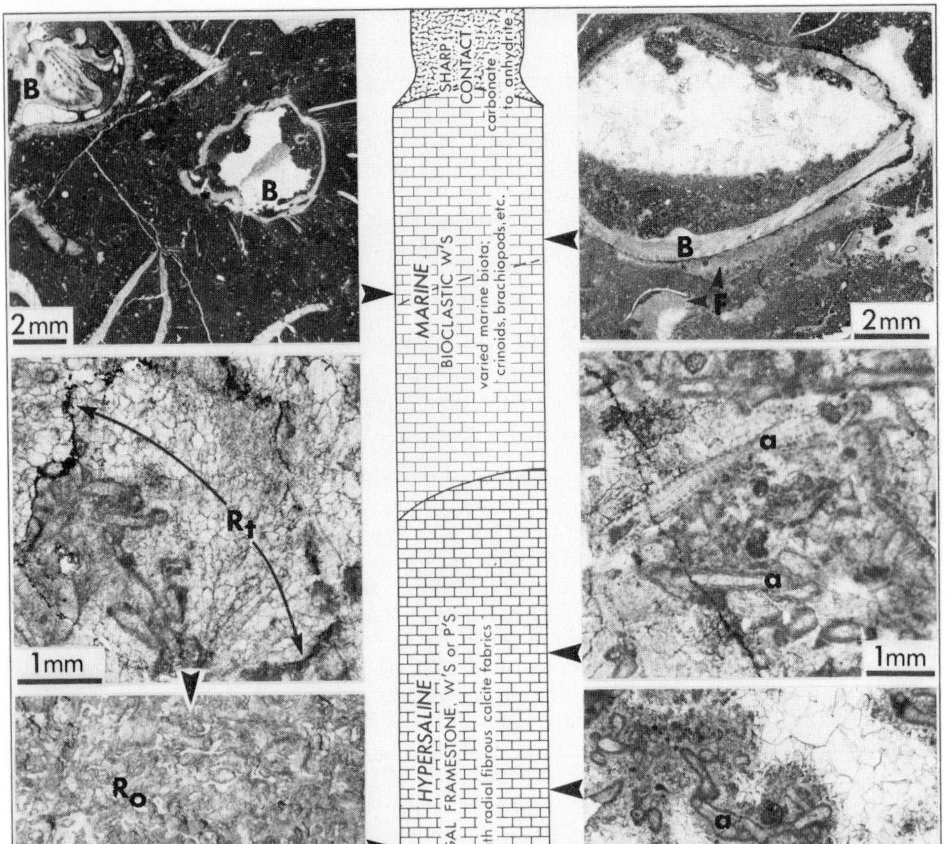
MARINE
BIOCLASTIC W'S
varied marine biota; crinoids, brachiopods, etc.
SHARP CONTACT
carbonate to anhydrite
HYPERSALINE
ALGAL FRAMESTONE, W'S or P'S
with radial fibrous calcite fabrics
less varied biota; ostracodes, spicules, calcispheres, algae
MARINE
CRINOID-RICH BIOCLASTIC W'S, P'S
varied marine biota; crinoids, brachiopods, ammonoids, bryozoans, etc.
SHARP CONTACT
anhydrite to carbonate
B
B
2mm
B
F
2mm
Rt
1mm
a
a
1mm
Ro
CM
a
a
1mm
B
A
C
1mm
C
C
1mm

FIG. 8–Generalized lithologic succession in an algal mound in the Otto Fiord Formation at van Hauen Pass (Fig. 1). Establishment of a crinoid-rich marine biota on an elevated, erosion-shaped substrate apparently was followed by hypersaline conditions favorable only to euryhaline algae and a few other organisms. Early diagenetic emplacement of radiating masses of aragonite reduced the primary and secondary porosity of the algal mound. Marine limestones that cap the algal mound accumulated during the transgressive phase of carbonate-evaporite cycles; anhydrite was deposited in off-mound areas. The uppermost carbonate-to-anhydrite contact is not interlaminated; it may be a marine hardground. Symbols: W'S–wackestone; P'S–packstone; C–crinoid; B–brachiopod; A–ammonoid; a–algae; F–fibrous calcite; Ro–radial-fibrous calcite, outcrop surface; Rt–radial-fibrous calcite, thin section.

2. The basal unit of the mound is formed by 2 to 5 m of well-bedded, dark-colored limestone, in part argillaceous and recessive; unit thins toward flanks of the mound, and apparently is continuous with the lowest and thickest of three correlative limestone beds in the intermound setting (Fig. 7). This basal limestone unit contains a rich marine biota dominated by crinoids; other common components are brachiopods, ammonoids, bryozoans, gastropods, ostracodes and foraminifers. Darker micritic lumps of sediment, perhaps fecal, are common. Matrix is bioturbated spicular, pelletoidal wackestone. Many skeletal clasts are bored. In beds higher in the unit, sediments were reworked by currents into poorly graded beds of crinoid grainstone and crinoid-bryozoan packstone with traces of quartz silt and sand. Uppermost beds include finer-grained, spicule- and calcisphere-rich bioturbated wackestone with large, thick-walled ostracodes and the lowermost algae.

3. The main topographic expression of the mounds is the result of buildup by branching tubular algae. The dominant varieties are beresellid and kamaenid algae that have some affinities with dasycladacean algae, but they may be unrelated to that family (Mamet and Rudloff, 1972). Preserved fabrics range from open framestones of algal mesh with patches of pelletoidal wackestone matrix, to aggregates of broken algal clasts in grainstone and packstone textures. Other skeletal components in the algal-mound facies are brachiopods, fenestellid bryozoans, crinoids, encrusting foraminifers, ostracodes, and several other types of algae including *Komia* (or *Ungdarella*). Internal hard grounds also may be present in these rocks. Where the primary skeletal frame is preserved, algal branches commonly are encrusted by a fringe of isopachous, inclusion-rich fibrous calcite. This calcite is similar to fibrous calcite cements in Pennsylvanian and Permian reef rocks higher in the Sverdrup succession that are interpreted to be of submarine origin, and probably neomorphic after magnesian calcite (Davies, 1977).

Another characteristic of the algal-mound facies is the predominance in some units of dark-colored fan-shaped masses of radial-fibrous calcite (Fig. 8, left; also Wardlaw and Christie, 1975, Figs. 10-13). On exposed rock faces, dark-colored radial fibrous calcite contrasts with light-colored pore-filling diagenetic calcite and dolomite, producing a distinctive mottled fabric (Fig. 8, left). In thin section, radial fibrous calcite is pseudomorphic after radiating cones of a crystalline precursor, which grew in primary cavities within the algal frame, in other cavities perhaps of solutional origin, and in part developed by replacement. The radial, fibrous calcite fabric is post-dated by zoned, iron-rich calcite spar and zoned dolomite crystals, and by pore-filling and replacive quartz (chalcedony), anhydrite, and fluorite.

Essentially identical pore-filling and replacive growths of botryoidal and spherulitic calcite have been found in Pennsylvanian and Lower Permian reef rocks higher in the Sverdrup succession; these fabrics are interpreted to be neomorphic after aragonite of submarine origin (Davies, 1977).

Large selenite crystals occur at several horizons within the algal-mound units. These crystals are interpreted to be very late diagenetic, post-dating unloading of the Otto Fiord Formation by erosion in Late Cretaceous and Tertiary time (see Davies, 1976, for summary of burial history and post-burial diagenesis of overlying rocks). However, distribution of selenite crystals may coincide with original sulfate-rich depositional discontinuities within the mound mass.

4. The algal-mound facies is overlain by several units 3 to 4 m thick of dark-colored, argillaceous skeletal wackestone that contains a varied marine biota. These beds drape across tops of the mounds and thin out at the flanks. Apparently they are continuous with the two thin limestone beds in the intermound setting (Fig. 7). Common skeletal clasts include crinoids, large brachiopods, sponges and sponge spicules, ammonoids, bryozoans, and foraminifers. The matrix contains abundant pelletoids and spicules. Whole sponges and brachiopod valves with delicate spines still intact and embedded in the matrix attest to the autochthonous nature of this marine biota and of the sediment. Irregular fenestrae and skeletal shelter pores in these rocks commonly are rimmed by fibrous calcite cement, apparently of submarine origin (Fig. 8). However, some fenestrae lack this cement lining; they may have been formed by vadose processes. Remaining primary pores in this facies are filled by diagenetic, zoned ferroan

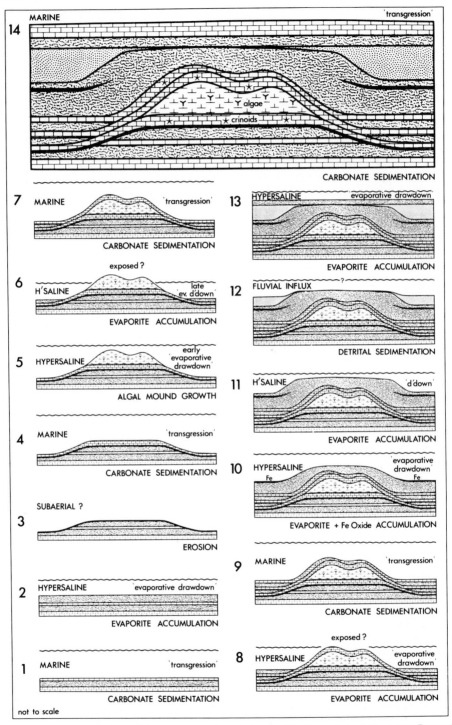

FIG. 9—Interpretation of the sequence of events involved in formation of the Lower Pennsylvanian algal mounds and associated lithologic units at van Hauen Pass. Stages 1 to 13 include at least four "transgressive-regressive" cycles that correlate with marine ingression followed by evaporative drawdown. Stage 14 records the beginning of a fifth cycle.

calcite and zoned dolomite. Significantly, radial fibrous calcite fabrics of the type that characterize the algal-mound facies are absent in these capping beds.

The uppermost bedding surface of the capping skeletal limestone is very dark colored and pyritic, and on it are many large brachiopods, articulated segments of crinoid columnals, ammonoids, and other clasts of marine organisms. This surface probably represents a marine hardground. It is overlain directly by anhydrite or locally by gypsum. This contact is sharp, and lacks the interlaminated limestone and anhydrite characteristic of limestone-anhydrite cycles in this formation.

Evolution of the Algal Mounds

The crinoid-algal mounds in the Otto Fiord Formation at van Hauen Pass are interpreted as having been formed by the following sequence (see Fig. 9):

1, 2, 3. Interruption of rhythmic deposition of marine limestones and gypsum by subaerial solution and erosion, leaving an irregular surface of elevated "platforms" of residual gypsum, and depressions 14 to 15 m deep. (The alternative explanation of submarine erosion cannot be dismissed, but it is discounted because strata of other cycles, which apparently were exposed periodically to normal marine water, do not show evidence of irregular erosion. However, influx of hyposaline water could have produced this erosional-solutional event in a subaqueous environment.)

4. Transgression onto the unconformity and establishment of a varied marine biota on all substrates. Clay and silt transported in the sea or reworked from the eroded surface were incorporated into lowermost beds of carbonate rock. Organic productivity and carbonate-sediment accumulation rates were higher on elevated substrates, probably through combination of advantageous photic levels, favorable supply of nutrients, and low turbidity. Crinoid communities colonized submarine highs and increased the supply of skeletal carbonate. In depressions, crinoids were less abundant and thinner beds of sediment accumulated.

5. The transgressive maximum with its normal-marine salinities and biota was followed by progressive increase in salinity and depression of regional sea level under evaporative drawdown (Maiklem, 1971). Controls for this rhythmic pattern probably were mainly tectonic, reflecting evolution of the riftlike Sverdrup depression. As salinities increased, the varied marine biota on the sea floor was succeeded initially by an increasingly restricted euryhaline biota of calcisphere-producing algae and ostracodes, and finally by beresellid and kamaenid algae. With the high nutrient supply of the hypersaline water mass, algae flourished and constructed the main mound buildup. The calcified algal mesh partly was cemented into a rigid framework by submarine cements.

Pervasive cementation and replacement in primary pores and sediments by spherulitic and botryoidal aragonite was confined to the algal-mound phase, suggesting strong contemporaneous environmental and substrate controls. This early diagenetic process may have resulted from the internal circulation of hypersaline water derived from the enclosing water and sediments. However, this process possibly was concurrent with precipitation of sulfate sediments, as hypersaline water from which gypsum has precipitated should be enriched in magnesium relative to calcium, a factor favorable to precipitation of aragonite at the expense of calcite (Berner, 1975).

6. With continued net evaporation, production of skeletal carbonate was inhibited, and eventually gypsum was precipitated as bedded crystal aggregates in the intermound areas and at least part way up flanks of the mounds. Possibly, at the evaporitic maximum, tops of the algal mounds were exposed, although there is no diagnostic evidence of such an event.

7. This evaporitic cycle was terminated by renewed marine ingression and deepening. Some gypsum deposits higher on flanks of mounds may have been eroded. The mound surface and the intermound substrates were repopulated by marine organisms. Organic productivity was greater on the mounds, and thicker sediments accumulated there. A thin bed of marine limestone was deposited in the intermound areas.

8. With progressive increase in salinity, gypsum again accumulated in the off-mound setting. Absence of an intermediate algal-mound component in this cycle, in contrast

with Event 5, may have been caused by differences in water depth, salinity, rates of sea-level change, or probably, a combination of these factors.

9. Event 7 was repeated, producing the second of the thin intermound limestone beds and another lenticular mound unit of marine limestone.

10. The carbonate subcycle of Event 9 was followed by deposition of thick-bedded gypsum that apparently draped over and buried the mounds. Thickness of the sulfate rock and topography of the mounds indicate that original water depth for this depositional cycle was more than 30 m.

After deposition of about 30 m of gypsum in the intermound areas and burial of the mounds by many meters of gypsum, 2 to 3 m of iron oxides and gypsum accumulated, later to be converted to hematite and hematite-stained anhydrite. This deposit was confined to intermound depressions, but a thin trace of hematite extended a short distance toward and above flanks of the mounds.

The source of the iron oxides is open to conjecture. Hematite staining of evaporite-associated basinal carbonates of the Silurian Niagara Group in the Michigan basin is attributed to early loss of detrital organic matter at the sediment-water interface, under conditions of very slow "basinal" sedimentation. This inhibited bacterial reduction of hydrous iron oxides during subsequent burial; later dehydration to red iron oxides completed the staining cycle (Mesolella et al, 1974).

In rocks of the Otto Fiord Formation, red iron-oxide staining also is found in limestones and argillaceous limestones interbedded with anhydrite, but these are minor compared with the discontinuous hematite bed of the intermound areas, discussed above. Wardlaw and Christie (1975, p. 166) suggested that ferruginous sandstones at the basin edge may have been a source of iron, and later of the overlying sandstones. Lack of detrital grains in the hematite rules against movement of strong traction currents into this area from a fluvial source. However, an estimated increase in opaque argillaceous and organic impurities in the hematitic anhydrite would support the inference of a density outflow of low-salinity, clay-rich water into and across the hypersaline water. Iron oxides and other iron complexes introduced in this stratum may have precipitated from the mixing zone between the two water masses, and may have been preserved preferentially on floors of the deep depressions.

11. Iron-oxide accumulation was accompanied and followed by precipitation of gypsum and by crystal growth, yet the irregular depositional topography of buried mounds and depressions was maintained.

12. A major influx of detrital sediment into the hypersaline basin filled the partly gypsum-filled depressions between the gypsum-covered mounds. Skeletal grains in sandstones indicate reworking of contemporaneous or older marine sediments at the margins of the basin, and large plant fragments in the sandstones indicate a source from a vegetated landmass. Sedimentary structures in the sandstones indicate transport from the north or northeast. Basinward progradation of the detrital wedge is indicated by coarsening-upward from shales and silts to medium-grained sandstones. The detrital sediments reduced submarine relief to a more-or-less planar surface, with a thin green shale overlying the slightly elevated areas above former mounds.

This pulse of detrital sediment is interpreted to represent a phase of renewed tectonic uplift and basinal subsidence, perhaps to the east and northeast, where similar but thicker sandstones are interbedded with anhydrite of the Otto Fiord Formation close to the eastern margin of the basin. Progressive but episodic deepening of the Sverdrup depression through this time, consistent with this interpretation, is recorded by thickening of successive anhydrite and limestone units, and by increase in marine biotic components in limestones higher in the section.

13. Levelling of the submarine topography by detrital sediments was followed by accumulation of a blanket of gypsum of more-or-less equal thickness across the entire area.

14. Termination of the evaporitic subcycle and initiation of the next major depositional cycle was marked by accumulation of a thick, continuous skeletal limestone unit. The limestone is slightly thicker over the buried mounds, indicating that slight topographic expression of the mounds still existed.

Anhydrite-To-Carbonate Facies Transition, Northeastern Hare Fiord

At the head of Hare Fiord on northwestern Ellesmere Island (Fig. 1, locality 2), a complex facies change from Otto Fiord evaporites to carbonates is exposed along the north wall of the fiord (Figs. 10, 11). The regional setting of this exposure, and the type of facies transition, indicate that this locality is at or very near the northwestern edge of the evaporite depositional "basin"; it represents an evaporite-to-carbonate shelf facies change, modified by local reef facies. The description of this facies change is based largely on aerial field photographs and on a section measured near the talus chute at the left-center of Figure 10.

The stratigraphic succession at this locality is composed of more than 300 m of conglomeratic red beds of the Borup Fiord Formation, which lie with angular unconformity on Cambrian rocks. The Borup Fiord red beds are overlain by 180 m of bedded carbonates that are considered to be a facies of the lower Otto Fiord Formation. A kilometer or so along strike, this bedded carbonate-platform unit terminates abruptly, and its place is taken by anhydrite and lensoid masses of rubbly-weathering limestone (Fig. 11). The nature of the abrupt change is difficult to decipher. Either it is a fault or a very abrupt facies change; if a fault, it may be synsedimentary, predating deposition of evaporites, as beds in overlying cycles of anhydrite and limestone extend across the break without disturbance. Another alternative is that lensoid masses of limestone enclosed in anhydrite are slump masses, separated from the bedded limestone by small listric faults. However, the geometry of these lenticular limestones (Fig. 11) suggests that they are more likely to be in-place mounds or reefs, similar to those in the Otto Fiord Formation at van Hauen Pass (Fig. 7).

Lenticular limestones illustrated in Figure 11 were traced from color slides; they appear to be stacked and stepped-back toward the former platform edge. About 6 km further along strike, where the Otto Fiord Formation is fully developed and overlain by normal basinal rocks of the Hare Fiord Formation, four or five prominent limestone units are interbedded with the anhydrite; these limestones can be traced along strike and correlated approximately with the four or five limestone mounds. Similar lenticular limestone units are interbedded in anhydrite higher in the Otto Fiord Formation, above the stacked mounds; these also are interpreted to be in-place mounds (Fig. 11). Both groups of mounds probably are constructed largely of beresellid and kamaenid algae similar to the mounds at van Hauen Pass, as correlative limestones along strike contain many of these algae, some in submarine-cemented framestone fabrics.

The platform-limestone unit is composed of well-bedded, dark-colored skeletal wackestones and dolomitic wackestones containing a shallow-marine biota. The most

FIG. 10—North wall of Hare Fiord east of Girty Creek (Fig. 1, locality 2), illustrating anhydrite and carbonate facies relations at or near the northwestern margin of the Otto Fiord evaporite "basin." Visible section is about 500 m thick. Interbedded anhydrite, limestone and shale of the Otto Fiord Formation at lower right grade to the left (westward) into carbonates. Tongues of limestone (1, 2, 3) from the facies transition are keyed to Figure 11. Note the steep depositional dips of the carbonate tongues, and within massive limestones of the overlying transgressive Nansen Formation.

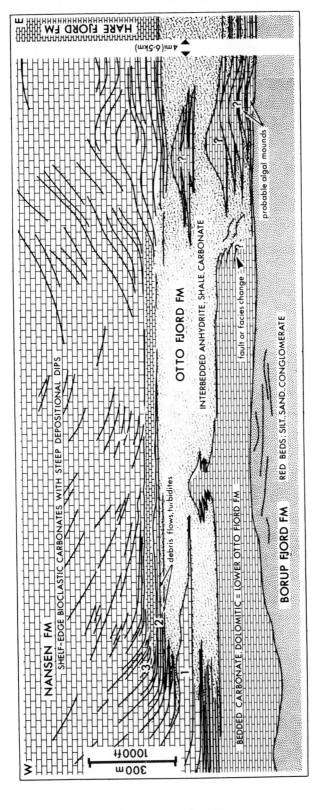

FIG. 11—Sketch of carbonate-anhydrite facies relations along the north wall of Hare Fiord east of Girty Creek (Fig. 1). The section is 2.5 km long; Figure 10 covers the western (left) one-third of this section. Lithologic column at right represents the more normal succession of the central basin, exposed about 6.5 km eastward. For regional perspective, this section can be compared with the western facies boundary of the Otto Fiord Formation shown in Figure 10. Limestone tongues 1, 2 and 3 are keyed to Figure 10.

common skeletal components are tubular and encrusting foraminifers and beresellid and other types of algae, with fewer crinoids, brachiopods, gastropods, ostracodes, bryzoans, fusulines, and calcispheres. A few beds contain ooids; most contain abundant micritic pelletoids. Several discontinuous beds of anhydrite up to several meters thick are interbedded with limestones of the platform unit.

Where it is continuous, the carbonate platform is overlain by about 180 m of interbedded anhydrite, limestone, and black, red, and green shales or argillaceous limestones of the more normal Otto Fiord section. The lower 50 to 80 m of this section grades westward into bedded limestones that appear to be of the same type as the underlying platform unit. Along strike to the east, the lower beds of anhydrite in the Otto Fiord interfinger with and surround another small limestone mass interpreted to be a reef (Fig. 11) with a composite thickness of about 80 m.

The 180 m-thick Otto Fiord section above the platform is interrupted about 100 m from the base by a prominent, discontinuous limestone bed that has a planar base and an irregular upper surface (Figs. 10, 11, unit 1). It varies from 40 to 50 m thick where it overlies the bedded anhydrite, but thins markedly to the east, so that 500 m along strike it is indistinguishable from other thin limestones interbedded in anhydrite. Westward, this irregular limestone stratum thickens, and within a few hundred meters can be traced into the base of a steeply dipping tongue of limestone that rises up depositional dip for at least several hundred meters before becoming unrecognizable. Apparent depositional relief on the upper surface of this steeply dipping limestone tongue and its subhorizontal extension into the anhydrite section is several hundred meters or more.

Limited sampling through the irregular limestone unit, where it is interbedded with anhydrite, indicates that beresellid algal-framestone fabrics with early fibrous cements are common in at least the central part of the unit. These fabrics are identical to those found in mounds of van Hauen Pass. The upper beds include pelletoidal skeletal packstones with algae as a common component.

The irregular limestone unit is overlain by well-bedded anhydrite, limestone, and red and green shaly strata of more normal Otto Fiord evaporite facies. Because of the irregular topography on the underlying carbonate, thickness of this evaporite section varies from 50 m to more than 100 m. Eastward, the evaporite section is continuous with the upper part of the fully developed Otto Fiord Formation, which is thicker than 300 m. Westward, however, a marked facies change is revealed where anhydrite beds pinch out between tongues of limestone that are extensions of steeply dipping shelf-foreslope or local reef limestones (Figs. 10, 11). Present dips on bedding planes in these foreslope rocks exceed 30°; extensive submarine cementation by fibrous cements appears to have been a controlling factor in stabilization of these steep depositional slopes. The abruptly thinning tongues of limestone that extend into the anhydrite section are characterized by beressellid algal-framestone fabrics with early fibrous-calcite cements. Depositional relief on the carbonate foreslope units is estimated to increase within each successive unit, and to be at least 400 m for the uppermost limestone-anhydrite cycles.

The evaporite section is terminated by several prominent, resistant limestones that are extensions of foreslope tongues, which thin out to the east (Figs. 10, 11, unit 2). The massive, discontinuous limestones above the evaporites include breccias and graded beds interpreted to be debris sheets and turbidites shed from the reef and shelf edge (Davies, in press). The most prominent of these debris beds is overlain by about 50 m of dark, recessive-weathering, argillaceous, crinoid-rich limestones that are typical of the "basinal" Hare Fiord Formation. These rocks grade westward into the same type of steeply dipping foreslope beds of the shelf edge (Fig. 11), with more than 450 m of depositional relief on the uppermost bedding planes.

This thin, dark limestone of the Hare Fiord is of deeper-water aspect, and is overlain abruptly by more than 1,000 m of light-colored limestone of the Nansen Formation. On field photographs, steeply dipping bedding planes can be followed throughout the Nansen Formation to delineate a complex of prograding foreslope-sediment tongues and local reeflike masses. However, these reeflike masses may be local spurs or promontories along the shelf-to-basin front, truncated randomly by erosion of the fiord wall. A few kilometers east of this area, massive shelf limestones of the Hare Fiord Formation in turn grade into basinal rocks of the Hare Fiord that are of normal thickness (500 m).

In summary, the exposure at locality 2 in the northeastern Hare Fiord reveals complexities that may be present at a shelf-to-trough transition from marine carbonates to "basinal" evaporites, and emphasizes the depths that water must have been during deposition of interbedded limestones. Significantly, this entire section records progressive deepening in successive transgressions into the Sverdrup depression. Relief on the lowermost carbonate units was only a few meters, but it increased to 300 m or more by the end of deposition of evaporites, and to 450 m or more during deposition of deep-water sediments of the Hare Fiord Formation. These events were followed by progradation of massive shelf-edge sediments into this marginal part of the basin floor.

Discussion
Significance of Facies Relationships for Evaporite Depositional Model

Facies relationships in the Otto Fiord Formation of evaporites and coeval or interbedded carbonates add support to the interpretation that the sulfate evaporites were deposited in a hypersaline subaqueous environment (Davies and Nassichuk, 1972, 1975; Wardlaw and Christie, 1975). The contrasting condition of subaerial exposure and diagenetic emplacement of sulfate in sabkha environments specifically is denied by the accumulated evidence.

If the interpretation of stages of deposition of crinoid-algal mounds at the van Hauen Pass locality is correct, limestones in the off-mound setting, which are coeval with marine limestones draped over the mounds, must have been deposited in at least 20 m of water. Similarly, if the thick anhydrite unit that overlies this uppermost marine limestone of the mound sequence was deposited during the "regressive," hypersaline phase of the "transgressive-regressive" cycle, water depth in the intermound areas at the beginning of accumulation of gypsum must have been at least 30 to 40 m, and perhaps considerably more. It should be stressed that these arguments for depths of water during accumulation of evaporites presuppose that bedding in the anhydrite units is related to seasonal or other nearly regular, periodic, climatic rhythms. These rhythms caused only minor perturbations in a progressive increase of salinity and decline of sea level; they do not allow for extreme fluctuations of sea level during deposition of strata composed of bedded anhydrite.

Estimates of water depth during deposition of limestone at the major facies change on eastern Hare Fiord are based on the differential relief shown by steep depositional dips on limestones interbedded with anhydrite (Figs. 10, 11). These depth estimates range from several meters at the base of the carbonate-anhydrite section, to more than 400 m for carbonate-anhydrite cycles at the top of the Otto Fiord section. Of course, the estimates apply only to the marine limestone, and not to anhydrite interbedded with the limestone; yet it is inconceivable and unnecessary that total desiccation occurred at the evaporitic maximum of each cycle. Even if desiccation did occur occasionally, the greater mass of evaporite minerals must have accumulated subaqueously prior to total desiccation.

Considering the Sverdrup basin on a regional scale, the anhydrite-limestone cycles exposed on northwestern Ellesmere Island probably are a shallower evaporite facies than the salt facies now buried in the central and southern regions of the basin. Although evaporites and other sedimentary strata may have been localized in sub-basins, the general conformity in age, lithologic type, and succession suggests interconnection of depocenters. Accumulation of individual strata of salt in the central and southern sub-basins is interpreted to have been coincident with basin-wide evaporitic maxima; thus, units of salt are correlative with the uppermost beds of anhydrite units in the northwest.

A corollary of the depositional model proposed for evaporites of the Otto Fiord Formation on northwestern Ellesmere Island is that coeval limestone units around margins of the evaporitic "basin" must have been exposed periodically as sea level was depressed during evaporitic maxima (Maiklem, 1971). This consequence is supported by the very shallow-water biotic composition of some of the limestones, by thin beds of oolitic and abraded-clast grainstones (Davies, 1975a), by diagenetic evidence for intensive cementation, recrystallization and replacement by calcite spar, and by episodes of solution and moldic filling of ooids and other grains, consistent with episodic exposure to freshwater vadose diagenetic processes.

Significance of Radial-Fibrous Cement Fabrics

The distinctive radial-fibrous calcite fabrics in the algal-mound facies at van Hauen Pass and in other mounds in the Otto Fiord Formation—as well as in some units more regularly bedded (Wardlaw and Christie, 1975, p. 160)—are interpreted to be neomorphic after radiating cones or botryoidal masses of aragonite. This interpretation is based mainly on comparison with similar but better-preserved fabrics in Middle Pennsylvanian bryozoan reefs and Lower Permian shelf-edge limestones higher in the Sverdrup succession on northwestern Ellesmere Island (Davies, 1975a, 1977). In these rocks, botryoidal and sperulitic calcite fabrics are neomorphic replacements of a crystalline precursor that precipitated in primary pores and early fracture cavities, and that grew by replacement of primary, marine carbonate sediment and very early submarine cements. These diverse growth forms are found in algal mounds of the Otto Fiord, but are differentiated less clearly.

Evidence of an aragonite precursor of botryoidal and spherulitic calcite in rocks above the Otto Fiord Formation includes the type of pseudomorphic crystal fabrics, style of diagenetic replacement, and retention of high strontium levels (8,000 ppm) in some very large botryoidal calcite masses, similar to the strontium content of modern, marine aragonite precipitates (Davies, 1977).

Accumulated evidence suggests that the aragonite precursor in Pennsylvanian bryozoan reefs grew within the reefs while they were in the marine environment, with the qualification that much of the more massive aragonite may have precipitated in submarine fracture cavities, deep within the internal mass of the reef. This evidence of an essentially submarine origin of the aragonite includes the isotopic composition of botryoidal calcite, its high strontium content, petrofabric relation, interlayering with marine internal sediment, and comparison with modern submarine analogs.

In algal mounds of the Otto Fiord Formation, more intense recrystallization and replacement prevents a similar analysis of compositional and petrofabric relations. However, distribution of the radial-fibrous precursor clearly was controlled by the primary pore system in the algal-mound facies; in the mounds it was confined to this facies, and it predated precipitation of calcite spar and dolomite, and all other diagenetic and post-burial phenomena in these rocks. From this evidence, and by analogy with the botryoidal calcite fabrics in the overlying bryozoan reefs, the radial-fibrous fabric is interpreted to have been emplaced as aragonite penecontemporaneously with deposition, or as a very early diagenetic event while the mound remained in a marine or hypersaline subaqueous setting.

The hydrochemical conditions that favored this extensive cementation by aragonite in the porous algal mound may have included an elevated magnesium-calcium ratio in the surrounding water mass and in the internal pore-water system of the porous reef, coincident with precipitation of $CaSO_4$ as gypsum in the off-mound setting. Increase in the magnesium-calcium ratio of water beyond normal marine levels apparently favors precipitation of aragonite at the expense of calcite (Berner, 1975).

Comparisons of Crinoid-Algal Mounds, Otto Fiord Formation

Concentration of crinoidal debris in basal beds of limestone mounds at van Hauen Pass, compared with dispersal of debris in laterally-equivalent intermound limestone beds, attests to the early and prolific colonization of the elevated erosional substrate by crinoids, and to their contribution to nucleation of mounds. Crinoids served a similar nucleating function in mounds and reefs of many other Paleozoic basins, particularly buildups of Silurian, Ordovician and Devonian age (see Heckel, 1974, for a literature review of Paleozoic and other reefs). In normal marine basins the crinoidal nucleus commonly is overlain by coral, stromatoporoid, or other marine reefal facies; in contrast, mounds of the Otto Fiord were built up by euryhaline algae tolerant of the increasingly hypersaline conditions of evaporative drawdown.

Reconstruction of depositional events for mounds of van Hauen Pass (Fig. 9) merits comparison with the depositional models considered by Mesolella et al (1974) for Niagaran (Silurian) pinnacle reefs of the Michigan basin. Significantly, the model preferred by Mesollela et al (1974, Fig. 15, Model III) for Niagaran reefs is alternate deposition of carbonates and evaporites (as in the Otto Fiord area), rather than contemporaneous deposition in on-mound and off-mound settings, respectively. Mesolella et al (1974) referred to this condition as "quasicontemporaneous deposition." Changes in sea level during the carbonate-to-evaporite depositional cycle is inferred for reefs in both the Sverdrup and Michigan basins.

Complexities of Carbonate-Evaporite Facies Relationships

One of the obvious lessons to be learned from examination of exposures described here of evaporites in the Otto Fiord Formation is that in this type of tectono-sedimentary setting, facies relations between evaporites and carbonates are complex. Many such complexities would not be expected in an evaporite succession of sabkha origin, where depositional relief and thicknesses of carbonate-anhydrite rhythms should not be great.

This lesson is applicable particularly to subsurface studies, where interpretations are limited by drilling and seismic information. Dramatic changes over short distances in types and thicknesses of rocks normally might be interpreted to have been caused by faulting, yet rocks of the Otto Fiord Formation demonstrate that such conditions may be due to major basin-margin facies transitions.

Porous carbonate or detrital facies with reservoir potential (for example, the discontinuous sandstones and algal mounds discussed above) may be very localized within this type of geological setting, reflecting the influence of a combination of tectonic and depositional controls.

Conclusions

Exposures of the Otto Fiord Formation on northwestern Ellesmere Island demonstrate complexities that may exist at facies boundaries between carbonate rocks and evaporites in a basin-margin setting, and in organic buildups within the evaporite succession. These complexities are consistent with the hypersaline, subaqueous depositional model inferred for evaporites of the Otto Fiord Formation; conversely, they are not consistent with the facies relations expected in a sabkha evaporite deposit.

Depositional relief of several hundred meters, and thus minimum water depths, can be estimated from the geometry of limestone units that tongue-out between some anhydrite units at the exposed major basin-margin facies change from shelf carbonates to "basinal" evaporites. Although these water depths coincide with deposition of limestone, and although the sea was below these levels during evaporitic maxima, the anhydrite strata undoubtedly were deposited subaqueously.

Because of their continuity and completeness of exposure, evaporites, associated carbonates, and other rocks of the Otto Fiord Formation provide several useful guides for modeling evaporite-reef relations and carbonate-evaporite facies transitions. Application of these models to studies of similar carbonate-evaporite successions in the subsurface may prove beneficial to exploration for hydrocarbons in other sedimentary basins.

References Cited

Berner, R. A., 1975, The role of magnesium in the crystal growth of calcite and aragonite from sea water: Geochim. et Cosmochim. Acta, v. 39, p. 489-504.

Davies, G. R., 1975a, Upper Paleozoic carbonates and evaporites in the Sverdrup basin, Canadian Arctic Archipelago: Canada Geol. Survey Paper 75-1B, p. 209-214.

—— 1975b, Hoodoo L-41: diapiric halite facies of the Otto Fiord Formation in the Sverdrup basin, Arctic Archipelago: Canada Geol. Survey Paper 75-1C, p. 23-29.

—— 1976, Bitumen in post-burial diagenetic calcite: Canada Geol. Survey Paper 76-1C, p. 107-114.

—— 1977, Former magnesian calcite and aragonite submarine cements in upper Paleozoic reefs of the Canadian Arctic: a summary: Geology, v. 5, p. 11-15.

—— (in press) Turbidites, debris sheets and truncation structures in upper Paleozoic deep-water carbonates of the Sverdrup basin, Arctic Archipelago: SEPM Spec. Pub. 25.

—— and W. W. Nassichuk, 1972, Upper Paleozoic evaporites in Sverdrup basin, Arctic Canada (abs): AAPG Bull., v. 56, p. 612.

—— —— 1975, Subaqueous evaporites of the Carboniferous Otto Fiord Formation, Canadian Arctic Archipelago: a summary: Geology, v. 3, p. 273-278.

Heckel, P. H., 1974, Carbonate buildups in the geologic record: a review, in L. F. Laporte, ed., Reefs in time and space: SEPM Spec. Pub. 18, p. 90-154.

Maiklem, W. R., 1971, Evaporative drawdown—a mechanism for water-level lowering and diagenesis in the Elk Point basin: Bull. Canadian Petroleum Geol., v. 19, p. 487-503.

Mamet, B., and B. Rudloff, 1972, Algues Carboniferes de la partie septentrionale de l'Amerique du Nord: Revue de Micropal., v. 15, p. 75-114.

Meneley, R. A., D. Henao, and R. K. Merritt, 1975, The northwest margin of the Sverdrup Basin, in C. J. Yorath, E. R. Parker, and D. J. Glass, eds., Canada's continental margins and offshore petroleum exploration: Canadian Soc. Petroleum Geologists, Mem. 4, p. 531-544.

Mesolella, K. J., et al, 1974, Cyclic deposition of Silurian carbonates and evaporites in Michigan basin: AAPG Bull., v. 58, p. 34-62.

Nassichuk, W. W., 1975, Carboniferous ammonoids and stratigraphy in the Canadian Arctic Archipelago: Canada Geol. Survey Bull. 237, 240 p.

Thorsteinsson, R., 1974, Carboniferous and Permian stratigraphy of Axel Heiberg and western Ellesmere Island, Canadian Arctic Archipelago: Canada Geol. Survey Bull. 224, 115 p.

Wardlaw, N. C., and D. L. Christie, 1975, Sulphates of submarine origin in Pennsylvanian Otto Fiord Formation of Canadian Arctic: Bull. Canadian Petroleum Geol., v. 23, p. 149-171.

An Evaporitic Lithofacies Continuum: Latest Miocene (Messinian) Deposits of Salemi Basin (Sicily) and a Modern Analog[1]

B. CHARLOTTE SCHREIBER[2], RAIMONDO CATALANO[3] and EDWARD SCHREIBER[4]

Abstract Environments that may result in deposition of evaporites occur in a sedimentary continuum which begins in the subaerial continental environment and extends into a hypersaline sea. The evaporitic sedimentary sequence of latest Miocene (Messinian) age, near Salemi, Sicily, contains carbonate rocks, argillites and gypsum that are compared with sediments in a modern salina at Salina Santa Margherita di Savoia, Italy. Fine-scale similarities between sediments of the salina and those of the section at Salemi provide a basis for interpretation of ancient evaporites. The Messinian evaporites developed in a small basin during the pan-Mediterranean salinity crisis; they contain a vertical facies succession that follows the facies association observed in the working salina. By using the horizontal sedimentary sequence in the salina as a model record of evaporation of an ancient hypersaline water body, it becomes possible to relate certain textures of gypsum sediments to their paleoenvironments of deposition and other textures to post-depositional alterations.

Introduction

Environmental analogs of evaporite facies have been studied in arid sea-margin areas such as the Persian Gulf and the Red Sea. However, sediments produced primarily in a supratidal setting adjacent to open-marine seas are not representative of all evaporitive facies known from the rock record. In fact, because we live in a geologically atypical time, noted for absence of hypersaline oceans, there are few depositional analogs today that fit some of the theoretical milieus appropriate for formation of ancient evaporites, as suggested by geological studies. Therefore, evaporites that develop in the subaqueous environment have not been studied extensively, perhaps because they exist only in small man-made salt ponds. Marine salinas (salt works, or salt ponds) provide excellent subaqueous hypersaline analogs. For the most part these have been considered only from the point of view of water chemistry and isotope fractionation (e.g. Fontes, 1965, 1966), halite morphology (Shearman, 1970), and formation of carbonate sediments (Horodyski and Vonder Haar, 1975). However, they have not been studied extensively for sedimentary textures of gypsum that are produced within them. Only a few parallels have been drawn between the facies of modern salinas and those in the geologic record (Kinsman, 1969; Schreiber and Kinsman, 1975).

Considerable progress has been made in comparisons of the supratidal and intertidal facies produced in modern environments and the evaporite record in Sicily (Hardie and Eugster, 1971; Nesteroff, 1973; Sturani, 1975). This paper is addressed to aspects of the subaqueous facies deposited in water bodies that may have been of significant lateral extent (i.e., true evaporite seas) and unknown depth.

In a recent study (Schreiber et al, 1976), it has been proposed that when a hypersaline water body yields evaporites, these deposits form in facies relations that may be tied not only to water chemistry but also to the dynamics and biota of the water. Such considerations are summarized in Figure 1.

[1] Manuscript received March 30, 1976; accepted June 9, 1976.
[2] Department of Earth and Environmental Sciences, Queens College (CUNY), Flushing, New York 11367.
[3] Instituto di Geologia, Universita di Palermo, Corso Tuköry, 131, Palermo, Italia.
[4] Department of Earth and Environmental Science, Queens College (CUNY), Flushing, New York 11367. Also Lamont-Doherty Geological Observatory, Columbia University, Palisades, New York 10964.
International Geological Correlation Program (IGCP), Messinian Correlation Project 96, Contribution No. 7, and Lamont-Doherty Geological Observatory Contribution No. 2572.
We wish to thank Dott. S. Bommarito for his aid in our field studies. To Walter Alvarez and William Ryan we extend our thanks for discussions and helpful comments. R. Catalano and B. C. Schreiber were supported in this study by the North Atlantic Treaty Organization Grant No. 0854 (Columbia University, New York).

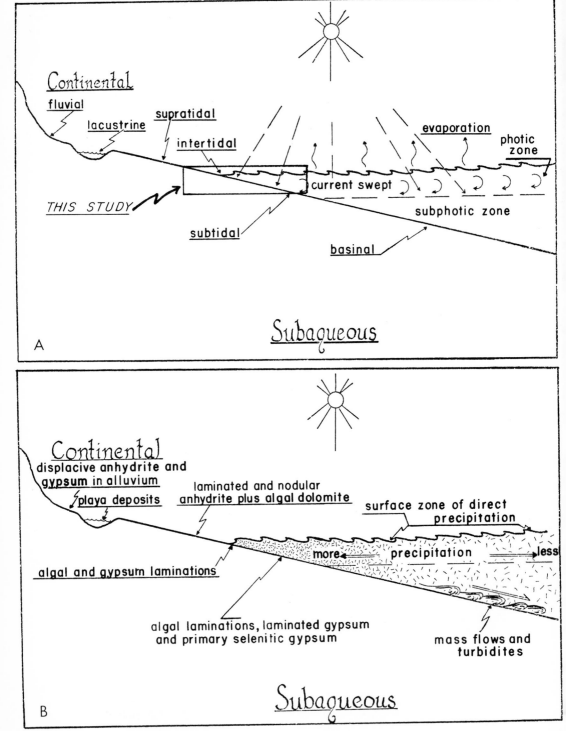

FIG. 1—Summary of the basic physical environments of evaporite deposition marginal to a hypersaline sea (A) and their sedimentary products (B). Environmental dynamics and biota in the water determine bedding, sedimentary structures, etc., while the physiochemical parameters control mineral species in each environment. (From Schreiber and Decima, in press).

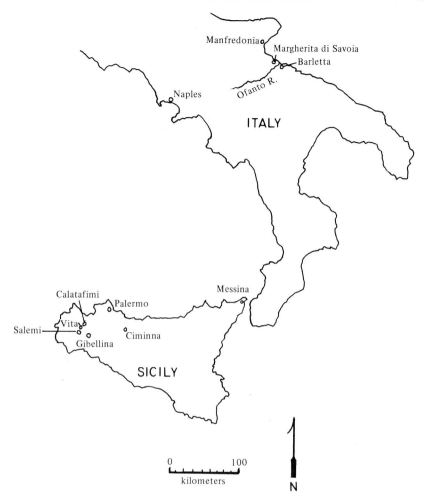

FIG. 2–The main localities cited in text, and nearby major cities.

This study presents a detailed description of the facies formed within related environments of deposition observed in a commercially operated salina. These facies, observed in a lateral sequence and a single time plane, are compared with those of a vertical evaporitic sequence at Salemi, Sicily (Fig. 2), in a latest Miocene (Messinian) deposit. Using the salina as an analog of changes in a hypersaline sea, inferences are drawn concerning the sequence in depositional environments through time as seen in the rock record.

Salina Santa Margherita di Savoia

The salina Santa Margherita di Savoia is located on the Adriatic coast of Italy between the towns of Manfredonia and Barletta (Fig. 2). Water from the Adriatic Sea enters the salina through a channel at the northern end of the saltworks and is concentrated in a series of ponds, where halite is deposited. The salt is harvested and the remaining fluid is processed for bromine and discharged into the sea, after 3 to 4 years in the salina.

FIG. 3—Top view of gypsum crust composed of prismatic crystals (1 to 2 cm long). Sample shown in photograph is from a hypersaline lagoon in the Persian Gulf (Azizyah).

The normal marine water that enters the first concentration ponds contains fish, which are gathered for human consumption. The fish subsist on the lush algal growth of these ponds and on the rich population of associated fauna. Numerous sea birds also feed on the smaller fish and crustacea and leave fecal residue that fertilizes the prodigious growth of algal complexes. The concentrating water is moved from pond to pond by a gentle gravity feed, controlled by barrier gates. Only at the entrance to the gypsum ponds and at the bromine-extraction station is the water pumped.

The bottoms and sides of the initial concentrating ponds are covered by diverse forms of algae. The next successive ponds, with higher salinity, are more restricted in flora and fauna. Although the water is saturated with respect to $CaCO_3$, carbonate does not appear to precipitate directly. However, samples of bottom sediment reveal numerous layers of algal mats (dead and living) that contain a great deal of $CaCO_3$ precipitate, as shown by strong reaction of the mats when tested with 5% HCl solution. Carbonate sediment occurs as fine, elongate needles within the organic mats. Samples from the mats putrified very rapidly and destroyed the carbonate; hence no X-ray study of the carbonate was made to determine which mineral phase was present.

In the concentrating areas just before the gypsum ponds, gypsum is deposited along margins of the water bodies as a crust of fine needles (Fig. 3). These crystals are acicular to prismatic, twinned and single crystals, arranged with long directions radially upward from the bedding surface. Growth of similar crystals under controlled conditions has been reported by Edinger (1973). Fine laminae of prismatic crystals, oriented with long axes within the bedding plane, accumulate in hollows and irregularities on the pond floors in the same area where acicular-crystal crust is formed. Some crystals

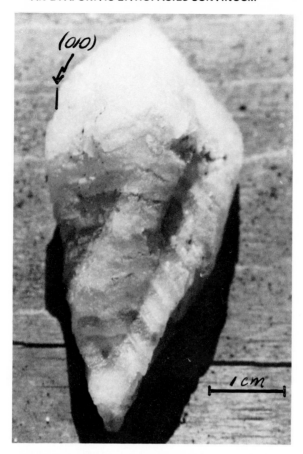

FIG. 4—Single gypsum crystal showing growth laminations on the (120) face and (010) cleavage in the vertical
direction (cleavage trace indicated by arrow). Sample collected at Salina Santa Margherita di Savoia.

in the laminae are broken, but many are doubly terminated. Similar doubly terminated
crystals float on the water at the outflow ends of the ponds.

Significant precipitation of gypsum begins in ponds where specific gravity of water
is about 1.133 (Baume grade 17). Gypsum appears as crusts of small crystals on floors
of the ponds and builds into beds of vertically oriented, laminated crystals (Fig. 4), or
crusts of fused crystals. The crusts develop into abutting polygonal slabs with large
clusters of rapidly growing gypsum crystals along the edges. The clusters of crystals are
called "cumulii" ("heaps") by workers in the salina. All the crystals are cone-shaped,
elongated perpendicular to the bedding surface, and laminated approximately parallel
to the bedding plane. Laminae are shaped as a "V" with apex upward; impurities are
concentrated within the laminae, which are (120) prism traces (following Palache et al,
1951, p. 482). The (010) cleavage trace is parallel to the long direction of the crystals.
Crystals having very similar appearance are shown by Fersman (1952) and by Schreiber
and Kinsman (1975); in these crystals, growth laminae are reported to be along the
(111) faces. Laminae within the crystals are 1 to 2 mm thick except in the cumulii
where each lamina is 3 to 4 mm thick. Approximately six to eight times per year,
less-concentrated water is added to each pond. Accordingly, six to eight laminae
develop in the 1 to 2 cm growth of crystals that occurs each year.

FIG. 5–Halite crystals developed around twigs. From the concentrated end of a gypsum pond at Salina Santa Margherita di Savoia.

The cumulii develop along pressure ridges or along margins of displaced segments of gypsum crust. Besides development of very thick laminae within crystals of the cumulii, broken segments of crust associated with the clusters also are displaced, uparched from the floor of the salina. A second generation of crystals develops beneath the displaced segments of crust. These new crystals are not laminated clearly, and commonly they form interlocking, irregular masses of tabular crystals. Where the tabular crystals come in contact, compromise crystal boundaries develop.

At the most saline end of the series of gypsum-forming ponds, crystal morphology of the gypsum is poor. A coarse centimetric mush of crystals develops, in which no particular orientation or form is observable. Numerous intermixed, small hopper crystals of halite also are present. Salt crystals also form on twigs and blades of grass (Fig. 5). Water in these ponds is red and contains a highly restricted and specialized biota, including many brine shrimp.

Salt developed in the salt-crystallization flats forms either at the air-water interface or at the sediment-water interface. Numerous hopper crystals are found both singly and in rafted masses in these ponds. Massive small overgrowths nucleate on all crystals that are on the bottoms of the ponds.

The Messinian Section at Salemi (Sicily)

The Messinian stratigraphic section is exposed in many areas of Sicily, in several large basinal deposits and several isolated smaller ones. The site of this study, one of the smaller basins, is in the west-central section of Sicily, near Salemi (Fig. 2). The stratigraphic section (Fig. 6) does not include halite, but ranges from limestone, marl,

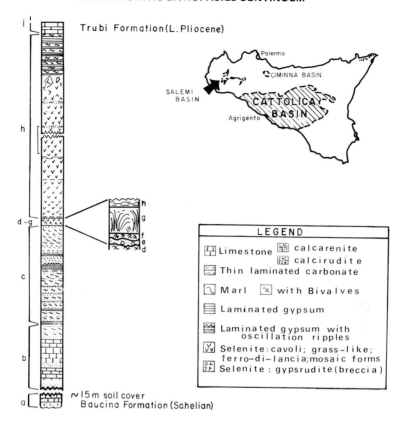

FIG. 6—Generalized section of the upper Miocene (Messinian) evaporite sequence at Salemi, Sicily. Location of surface outcropping of Messinian evaporite deposits is shown in inset.

and gypsiferous marl to gypsum. Absence of halite probably indicates that halite saturation never was reached, as no dissolution breccias or similar features are found. The section is capped by the Trubi Formation, lower Pliocene. The Messinian sequence, about 71 m thick, is exposed in vertical entirety, with only a little soil cover, in a section 3 km long that dips steeply into the hillsides. The evaporite facies observed in this area are rather different from the sequence studied in the much larger Cattolica basin (Fig. 6) by Ogniben (1957), Decima and Wezel (1971, 1973), and Schreiber et al (1976), but are more nearly the same as that seen in parts of the Ciminna basin (Fig. 2). Bommarito and Catalano (1973) examined the Ciminna section in considerable detail but did not describe the morphology of the facies. The facies of Vita, Calatafimi and Gibellina (Fig. 2), nearby small evaporite basins of the same age, have similar sedimentary sequences but differ in details.

In all of these Messinian evaporite sequences, the underlying and overlying sediments are very similar. Open-marine Tortonian marls normally are below the evaporites, and they are underlain in turn by a limestone facies. Overlying the evaporites is the white to pale tan Trubi foraminiferal ooze of the Lower Pliocene (Fig. 6). In places an unfossiliferous clastic bed or series of beds lies between the evaporites and the Trubi, termed the "Arenazzolo." At Salemi, the underlying limestone of the Baucina Formation (Fig. 6, a) primarily is a biocalcarenite, containing a normal, shallow-water marine fauna with some ooids. This is the same limestone as the "Formazione calcareoarenacea di Baucina" of Aruta and Buccheri (1971) and it underlies many but not all of the evaporitic sequences of Sicily—as for example at Ciminna (Bommarito and Catalano, 1973). These limestones are considered to be "Sahelian" or lower Messinian in age.

At Salemi the limestone is extensively cross-bedded and contains some ripple structures. In some places mollusca are in life position. The upward transition between the limestone and the evaporite section is exposed imperfectly; it is covered partially by soil and talus, but can be seen as thin-bedded calcarenites and calcirudites interbedded with marls. The fauna in these beds is gradually impoverished upward. The uppermost carbonate layers are very thinly bedded, millimetric dolomitic limestones that contain small blebs or granules of gypsum, forming a birdseye texture. Some of the limestones in the upper series exhibit mud cracks and mud-chip accumulations. The uppermost few marl beds contain so much gypsum that they may properly be termed marly gypsarenites. At the top of the carbonate section is an argillaceous arenite, in which a brackish-water pelecypod fauna is present (Fig. 6, b).

Above the pelecypod bed lies a section of laminated gypsum 4 m thick (Fig. 6, c), in which strong cross bedding and well-developed oscillation ripples are evident (Fig. 7). Examination of these laminae in thin section reveals that the original particles were fine prismatic gypsum needles and cleavage fragments, mixed with micrite. Now they are extensively but not completely recrystallized to an interlocking mosaic with sporadic reverse-graded bedding. Above this laminated gypsum are gypsarenites and argillaceous arenites interbedded with some pelecypod-bearing marly arenites and marls barren of fossils. This part of the section ends in an irregularly bedded, yellow stratum having a well-developed soil profile that contains plant roots and a considerable amount of lignitic material (Fig. 6, e). In hollows on the irregular upper surface of this bed are well-developed, vertical, elongate gypsum crystals (Fig. 6, f). The crystals are from 6 to 16 cm long, cone-shaped and clearly are laminated at 1 to 2-mm intervals (Fig. 8).

Immediately above the level of the individual crystals is a bed about 1 m thick (Fig. 6, g) that contains large cumulii of gypsum crystals (30 to 80 cm long) (Fig. 9). Crystals within the cumulii also are laminated, the laminae being somewhat thicker than those in crystals of the underlying layer. Above the bed of cumulii crystals is a 43-m sequence of beds of selenitic gypsum ranging in thickness from 0.5 to 2.5 m, and

FIG. 7—Slab from bed showing oscillation ripples. Laminae are composed of alternations of gypsum (gray) and carbonate (white). Sample collected at Salemi, Sicily.

separated by thin layers of gypsiferous calcarenites or gypsiferous marls (Fig. 6, h). The selenitic gypsum has varied morphology, each type of which is distinctive. These include the "ferro-di-lancia" (spears of iron) form of Mottura (1871-1872) and Ogniben (1957), the "cavoli" (cabbage) and the grasslike forms of Richter-Bernburg (1973) (also Schreiber et al, in 1976). In addition some gypsum beds are composed of large (3 to 8 cm) irregular crystals forming an interlocking mosaic (Garrison et al, in press). One bed near the top of the sequence is composed of a coarse (5 to 10 cm) angular gypsrudite (Schreiber et al). Many of the gypsum crystals—in all of these categories— show millimetric laminations marked by micrite, pelletal carbonate, and organic inclusions. At the top of this evaporite sequence is an abrupt contact with normal marine rocks, represented by the lowest Pliocene Trubi carbonates (Fig. 6, i).

Comparison of Deposits in Salinas with the Section at Salemi

Marine salt ponds produce a horizontal sequence beginning with deposition of layered algal mats and associated carbonate, passing to crystallization of gypsum and then salt, but the morphology of the precipitates and the sedimentary textures produced vary in detail. These facies are products of an increasing salinity gradient. Gypsum crystals observed in the Salina Santa Margherita di Savoia include (a) prismatic needles that form at the water surface and that are deposited as detrital gypsum laminae, (b) prismatic needles that form on the pond floors; (c) individual cone-shaped crystals formed at the pond floor with vertical crystal orientation and horizontal laminations, and (d) large clusters of cone-shaped crystals, the cumulii, which also are laminated. The laminae develop as a consequence of refreshment of the pond with less-concentrated waters. This variety of crystal forms of gypsum differs from that observed in other salinas, despite the fact that all are shallow-water, hypersaline marine environments (e.g. Schreiber and Kinsman, 1975).

Differences may arise because concentration is not the only factor in development of the evaporitic sediment; biogenic factors may also be involved. For example, studies by Davis (1974) indicate that for rapid evaporation and the ensuing concentration of salt to continue, a delicately balanced life assemblage of algae, protozoa, and bacteria (particularly the halophilic forms) needs to be present in the concentrating ponds. This biotic assemblage causes red coloration of the water. Without a flourishing population

1 cm

FIG. 8–Laminated gypsum crystal showing the natural, weathered surface. Sample collected at Salemi, Sicily.

FIG. 9–Broken surface showing cumulii of crystalline gypsum. Laminations are not evident on this scale. Hammer is 45 cm long. Sampled at Salemi, Sicily.

of this sort, evaporation slows and may even cease. This occurs for two reasons: first, the higher the salinity of water, the greater the energy needed to cause evaporation; second, the higher the salinity of water, the greater the surface tension of water. The red color makes the salina a more efficient absorber of radiant energy, thus raising the water temperature above that of surrounding air, and thereby increasing evaporation. The biota also control the surface tension of water, which affects the evaporation rate. These factors are so important that if a pond had an inappropriate biota, or were to be sterile, it would not produce salt on an economic basis.

Gypsum that constitutes the Messinian evaporites of the Salemi area of Sicily includes all the forms developed synchronously along a salinity gradient in the Salina Santa Margherita di Savoia, except for salt. In the salt ponds, sand-sized prismatic gypsum, as whole crystals and cleavage fragments, is reworked and deposited as laminae on the pond floors. These are analogs of the laminated gypsum beds of Salemi. The cone-shaped crystals and cumulii of the salina also are direct analogs to crystals and cumulii of Salemi (cf. Figs. 4 and 8). In addition, several other forms are noted at Salemi. Of those remaining, the cavoli (cabbage) form is found at another present-day salina, the Leslie Salt Works at Newark, California (Schreiber and Kinsman, 1975; Schreiber et al, in press).

The other morphologic forms seen at Salemi, the grasslike and ferro-di-lancia (spears of iron) forms, the interlocking mosaic, and the angular gypsrudite have not been observed in a salina. However, within all of these forms impurities are observed that are suggestive of shallow-water formation: these include ooids, carbonate pellets, and numerous organic fragments. On floors of ponds in modern salinas, comparable accumulations of miscellaneous material are incorporated into gypsum. The pellets are brine-shrimp feces and the organic material is a mixture of algal-mat fragments, brine-shrimp eggs, and bits and pieces of nearby plants (reeds and pond-margin weeds).

The interlocking mosaic and gypsrudite forms have been studied in some detail in other Messinian deposits. From studies by Schreiber and Decima (in press) and Garrison et al (in press) it appears that the mosaic represents selenite crystals exposed to undersaturated water for prolonged periods of time, producing a dissolution residue of gypsum. Subsequent reconcentration of the water results in renewed crystal growth on the upper surface, and precipitation of gypsum cement and overgrowths between and around the residual selenite fragments below. This form is quite different in appearance from the angular gypsum rudite that represents accumulation of mechanically broken crystal fragments, with very little cement or overgrowth (Parea and Ricci-Lucchi, 1972; Schreiber et al 1976). Naturally, a mechanical breccia could become a residual deposit if the overlying waters were refreshed for a prolonged period of time. Water depth at which these forms develop is not determined.

Bedding structures and the relationship of the associated intercalated carbonates, argillites and laminated gypsum observed at Salemi are consistent with the conclusion that they developed in shallow water—and more particularly so when these features are compared with sedimentary products of the analogous salt ponds. Corroborative evidence of a shallow-water environment is found in the presence of oscillation ripples, together with such high-energy features as cross bedding and climbing ripples. The mud cracks and mud-chip accumulations, and soil development testify to periodic desiccation. Pelecypods (normally found in estuaries and brackish-water lagoons) imply nearby sources of fresh water, from land. These features, taken together, lead to the inference that in Salemi, there existed a shallow-water, restricted, hypersaline basin.

Composition of the sediments and the associated crystal morphology presented here from Salemi records, in part, the depositional sequence obtained by a synchronous lateral transect of ponds in a salina. Moreover, by splicing this sequence into that of the salina at Newark, California, an even broader approximation of the facies panoply that develops in an uncontrolled natural sedimentary process is created. The lateral succession of evaporite facies obtained from such a composite salina is, in a sense, a time sequence in a hypersaline ocean with a hydrologic deficit. The facies comparison of the lateral succession of deposition in a modern salina to a vertical sequence in the Messinian rock record, is an uncommon application of Walther's Law (1893-1894), which we consider to be useful in unraveling the historical record of the earth.

Although not all of the basic crystal forms of gypsum in the Messinian evaporites at Salemi have been observed to form in salinas, it should be noted that the number of marine salinas studied from this point of view is small, and the forms observed in the different salinas were not all the same. The many salinas that remain to be studied leave considerable opportunity for further discovery.

References Cited

Aruta, L., and G. Buccheri, 1971, Il Miocene preevaporitico in facies carbonatico-detritica dei dintorni di Baucina, Ciminna, Ventimiglia di Sicilia, Calatafimi (Sicilia): Rivista Mineraria Siciliana, v. 22, p. 188-194.

Bommarito, S., and R. Catalano, 1973, Facies analysis of an evaporitic sequence near Ciminna (Palermo, Sicily), in C. W. Drooger, ed., Messinian events in the Mediterranean: North Holland Publishing Co., Amsterdam, p. 172-177.

Davis, J. S., 1973, The role of micro-organisms in the production of solar salt, in A. H. Coogan, ed., 4th Symposium on Salt, Northern Ohio Geol. Soc., v. 2, p. 369-372.

Decima, A., and F. C. Wezel, 1971, Osservazioni sulle evaporiti messiniane delle Sicilia centro meridionale: Rivista Mineraria Siciliana, v. 22, p. 172-187.

—— —— 1973, Late Miocene evaporites of the central Sicilian basin, in W. B. F. Ryan and K. J. Hsu, et al, Initial reports of the Deep Sea Drilling Project, Leg XIII: U.S. Govt. Printing Office, Washington, D.C., p. 1234-1240.

Edinger, S. E., 1973, The growth of gypsum: Jour. Crystal Growth, v. 18, p. 217-224.

Fersman, A. E., 1952, Geological-mineralogical investigations of the Sakskoe Lake: Izbrannye trudy, Moskva, Izdatel'stvo AN SSSR, v. 1, p. 809-822.

Fontes, J. C., 1965, Fractionnement isotopique dans l'eau de crystallisation du sulfate de calcium: Geol. Rundschau, v. 55, p. 172-178.

—— 1966, Intérêt en géologie d'une étude isotopique d l'evaporation; cas de l'eau de mer: Acad. Sci., C. R., Sér. D., v. 263, p. 1950-1953.

Garrison, R., et al, (in press) Structures and petrology of Messinian evaporites cored during leg 42A (371, 372, 374, 375, 376, and 378), *in* K. J. Hsu, and L. L. Montadert, et al, Initial Reports of the Deep Sea Drilling Project.

Hardie, L. A. and H. P. Eugster, 1971, Depositional environment of marine evaporites– shallow, clastic accumulation: Sedimentology, v. 16, p. 187-220.

Horodyski, R. J., and S. P. Vonder Haar, 1975, Recent calcareous stromatolites from Laguna Mormona (Baja, California): Jour. Sed. Petrology, v. 45, p. 894-906.

Kinsman, D. J. J., 1969, Modes of formation, sedimentary associations and diagnostic features of shallow-water and supratidal evaporites, *in* Evaporites and petroleum: AAPG Bull., v. 53, p. 830-840.

Mottura, S., 1871-1872, Sulla formazione Terziaria nella zona zolfifera della Sicilia: Mem. R. Comit. Geol. Italia, n. 1, p. 50-140.

Nesteroff, W. D., 1973, Petrographie des evaporites messiniennes de la Mediterranee: Comparison des forages JOIDES-DSDP et des depots du Bassin de Sicile, *in* C. W. Drooger, ed., Messinian events in the Mediterranean: North Holland Publishing Co., Amsterdam, p. 111-123.

Ogniben, L., 1957, Petrographia della Serie solfifera siciliana e considerazione geologiche relative: Mem. Descr. Carta Geol. Italia, v. 33, 275 p.

Palache, C., H. Berman, and C. Frondel, 1951, Halides, nitrates, borates, carbonates, sulfates, phosphates, arsenates, tungstates, molybdates, etc., *in* The system of mineralogy of James Dwight Dana and Edward Salisbury Dana, Yale University, 1837-1892: John Wiley & Sons, 7th ed., rev., 1124 p.

Parea, G. C. and F. Ricci-Lucchi, 1972, Resedimented evaporites in the Periadriatic Trough (Upper Miocene, Italy): Israel Jour. Earth-sciences, v. 21, p. 125-141.

Richter-Bernberg, G., 1973, Facies and paleogeography of the Messinian evaporites in Sicily, *in* C. W. Drooger, ed., Messinian events in the Mediterranean: North Holland Publishing Co., Amsterdam, p. 124-141.

Schreiber, B. C. and A. Decima, (in press), Sedimentary facies produced under evaporitic environments, a review: Rivista Mineraria Siciliana.

—— et al, 1976, The depositional environments of the Upper Miocene (Messinian) evaporite deposits of the Sicilian basin: Sedimentology, v. 23, p. 729-760.

—— D. J. J. Kinsman, 1975, New observations on the Pleistocene evaporites of Montallegro, Sicily and a modern analog: Jour. Sed. Petrology, v. 79, p. 469-479.

Shearman, D. J., 1970, Recent halite rock, Baja, California, Mexico: Inst. Mining Metallurgy, Trans., v. 79, p. B155-B162.

Sturani, W. D., 1975, Relazione relativa al contratto di ricerca 73.01036.05. Il significato della crisi di salinita del Mediterraneo: C. N. R., Turin, p. 50-63.

Vai, G. B., and F. Ricci-Lucchi, 1977, Algal crusts, autochthonous and clastic gypsum in a cannibalistic evaporite basin—a case history from the Messinian of the northern Apennines: Sedimentology, v. 24, p. 211-244.

Walther, J., 1893-1894, Einleitung in die Geologie als historische Wissenschaft: Fischer Verlag Jena, 3 vols., 1055 p.

Base-Metal Concentration in a Density-Stratified Evaporite Pan[1]

P. SONNENFELD[2], P. P. HUDEC[2], A. TUREK[2], and J. A. BOON[2]

Abstract In a density-stratified lagoon, an algal mat flourishing in the photozone below the interface between surface waters and a hypersaline brine is being replaced by gypsum; waters beneath the photozone are anaerobic and deposit only organic oozes. Major- and trace-element concentrations in the brines increase with depth. Both gypsum and organic matter extract significant quantities of trace metals from the water. A layered, solar-heated brine is suggested as the mechanism for base-metal concentration in depositional environments of euxinic shale.

Introduction

The Los Roques atoll is situated about 145 km north of Caracas, Venezuela. In its northeastern corner lies the island of Gran Roque (Fig. 1) consisting of a ridge composed of a complex assemblage of Tertiary basic and acidic igneous rocks and a skirt of flat-lying fine-grained sediments. These sediments are mainly erosional products of the hill chain, and they form a thin veneer on a Pleistocene fringing reef. A reef barrier is growing around part of the island.

Annual rainfall at the nearest weather station, located about 60 km to the east, is about 15 cm and the annual rate of evaporation is 2.26 m (Schubert, 1974). The net deficit in precipitation over evaporation therefore is 2.1 m/yr. This results in very sparse vegetation over most of Gran Roque island, with mangroves restricted to narrow strips along the eastern, windward coast.

An enclosed lagoon on Gran Roque, called Lago Pueblo (Fig. 1), shows periodic density stratification and consequent heliothermal temperature regime (Hudec and Sonnenfeld, 1974), typical of density-stratified or meromictic lakes in semiarid and arid climates. Dense brine is generated as a direct result of excessive evaporation. Periodic runoff from the nearby hills and seawater swept in from the northeast by occasional storms provide an upper layer of low-salinity water. Because of the high density of the brine, seawater is considered not to percolate into the subsurface beneath the lagoon.

Precipitation of Gypsum

The surface layer of seawater evaporates slowly, becomes more saline, mixes only gradually with the underlying waters, and creates a slowly widening transition zone (Fig. 2). This process has been duplicated successfully in an artificial tank; the interfaces remain visible for a long time. Before mixing is complete, a new supply of low-salinity waters by rain or inflow creates a new surface zone. All the water below the new fresh water now becomes "brine." The major part of the solar energy absorbed in the surface layers is given off through evaporation. Solar radiation reaching the lower stratified layers is trapped, because convective currents do not cross the chemocline, i.e., the interface between waters of different salinity. The resulting heating establishes a thermocline; a $20°$ to $25°C$ temperature differential exists between low-salinity surface waters and the waters beneath. Calculations indicate that more than 90% of the solar energy reaching the system is absorbed within the uppermost 1 m of water.

Most of the lagoon is shallow, water depths varying seasonally between 60 and 100 cm. An algal mat flourishes within the photozone. Pure gypsum is precipitated; it develops typical pressure ridges and a cuirass of upright crystals similar to dogtooth spar on the exposed surface. The gypsum appears to replace aragonite precipitated by the algae, but stromatolitic laminations nonetheless remain preserved within the mat (Fig. 3).

[1] Manuscript received December 4, 1975; accepted June 9, 1976.
[2] University of Windsor, Windsor, Ontario N9B 3P4, Canada.

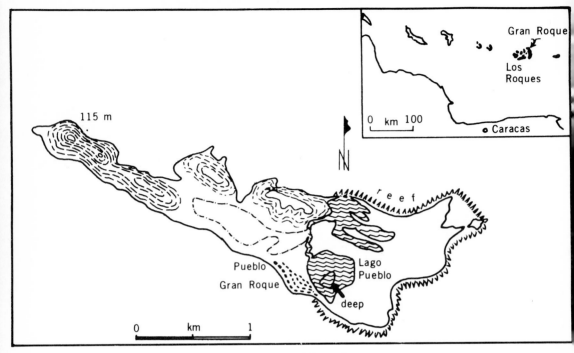

FIG. 1–Location of the study area.

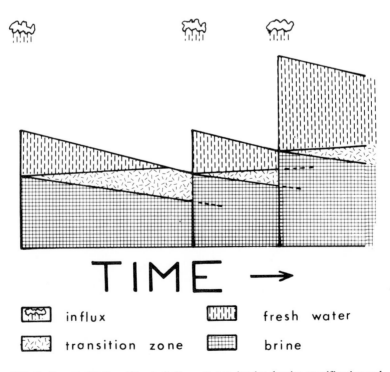

FIG. 2–Repeated influx of low-salinity waters maintains density stratification and
a consequent hot zone beneath chemocline.

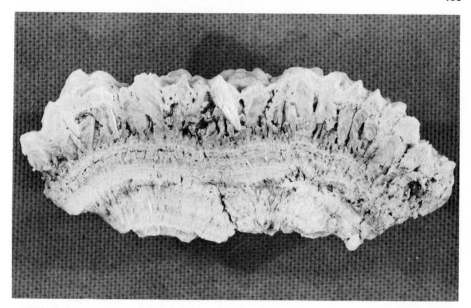

FIG. 3—Cross section of the algal mat, showing the outside surface covered by large, bladed crystals of gypsum. Layering is distinguishable by grain size and orientation, and in fresh samples by colors characteristic of each algal layer. Aragonite content is too small to be detected by X-ray diffraction.

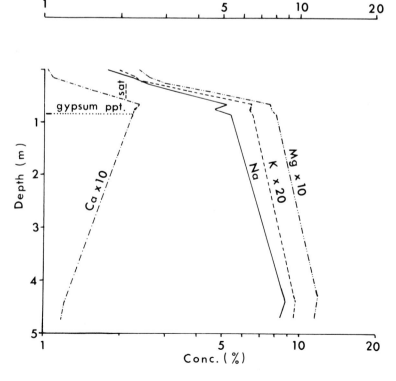

FIG. 4—Concentration of Na, K, Mg, and Ca in waters of Lago Pueblo.

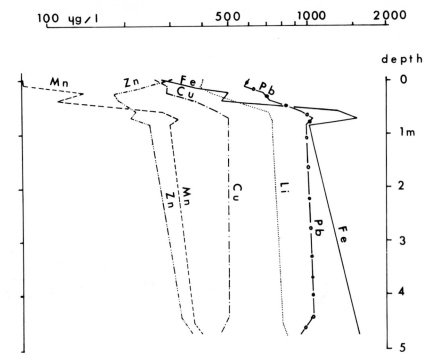

FIG. 5—Trace-element concentration in waters of Lago Pueblo.

Near the southwest corner of the lagoon, a small portion is as deep as 5 m, with vertical walls formed by overhang of the shallow-water gypsum mat. At depths more than about 1.5 m below the water surface—i.e. below the photozone—this deeper part is lined with black organic muds. Some comminuted gypsum swept over the rim of the shallow part of the lagoon has fallen into the deep mud. Density stratification prevents overturn or oxygenation of the waters and the mud is anaerobic. It contains significant quantities of hydrogen sulfide, probably in part derived from disintegration of gypsum. If so, the gypsum may have been reduced by anaerobic bacteria.

The most likely process of replacement of aragonite by gypsum is believed to be as follows: CO_2 absorbed at the air-water interface is inhibited by density stratification from penetrating freely into the hot zone, where algae live. This introduces fierce competition for available CO_2. As hydrogen sulfide rises from the anaerobic zone, it comes into contact with oxygen produced by algae just beneath the chemocline, and is converted to sulfate ions. The sulfate ion attacks precipitated aragonite, replaces it with gypsum, and produces carbonate and bicarbonate ions. These ions also are inhibited from penetrating the chemocline, and are used by the algae to build organic matter. Carbon thus is recycled from precipitated aragonite to bicarbonate and carbonate ions to organic carbon or new aragonite precipitate.

Concentration of Elements in Brine

Chemical analyses of the waters of Lago Pueblo using atomic-absorption spectrophotometry are shown graphically in Figure 4. Gypsum is precipitated beneath the thermocline, in the zone of hot water, where brine has reached a considerable degree of supersaturation; anhydrite has not been detected by X-ray diffraction. Concentration of brine continues until, in the deeper part, amounts of Na, K, Mg, Li, and Mn are about 4.8 to 4.9 times the amounts at the surface. Surface waters of the lagoon are not necessarily equivalent chemically to seawater surrounding the island. Calcium, carbon, and sulfur are extracted by precipitation of gypsum and aragonite. Magnesium is trapped in some magnesium-rich calcite in beach rock. Sodium and chlorine are lost through semiannual precipitation of halite on the windward shore (more than 100 sacks of rock salt are exported each April and August). Potassium is adsorbed in small quantities by bottom oozes.

Iron, copper, and lead (but not zinc) are concentrated much more than 4.9 times between the surface and the lagoon floor (Fig. 5). Some of the iron and other metals also are incorporated in the various layers of the gypsum crust. Comparison of surface waters with waters below the chemocline, at the level where gypsum is precipitated and with the gypsum mat itself (Fig. 6), shows gross deficiency in lead, zinc, and copper in waters in contact with gypsum. In part, this deficiency may be caused by absorption by macerated organic matter in suspension in the hypersaline brine. Manganese can be accounted for almost completely by a small amount incorporated in the crust, and a larger amount in solution. In contrast, iron shows enrichment compared to its equivalent concentration in surface waters.

Figure 6 shows graphically the metal content at the level of gypsum precipitation. Whereas, at this concentration gypsum precipitates, metal content is taken into consideration in the proportional amount of gypsum formed. This ratio has been taken as the average value of all gypsum analyses. Metal content of the brine as compared to the metal content of the surface water is considered next. If equilibrium conditions existed, the percentage of metal in the brine and in gypsum should add to 100; this is the case only for manganese. All other metals show either excess or deficiency from the norm, for reasons already described.

Content of Metals in Gypsum and Bottom Muds

The gypsum crust shows wide ranges of trace-element distribution from sample to sample. Different layers of the gypsum crust were sampled, and the dissolved samples were then analyzed by atomic-absorption spectrophotometry.

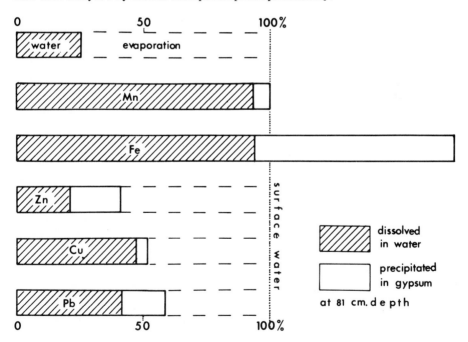

FIG. 6—Trace-metals balance in water at the level of gypsum precipitation. Top bar represents degree of concentration of brine, relative to seawater.

Table 1. Trace-element Content (in parts per million) of Gypsum from Gypsum Crusts and of Organic Matter from Bottom Oozes. Coefficients of correlation refer to amounts of metals and organic matter in samples of bottom mud.

	Gypsum		Bottom Muds	
Element	Range	Mean	Organic Matter	Coefficient of Correlation
Mn	3-5	4.0	159	0.934
Fe	40-492	145.9	35,184	0.919
Zn	13-101	39.2	10,147	0.807
Cu	4-10	7.4	859	0.946
Pb	52-89	69.7	787	0.991

Manganese and lead seemingly are distributed within a 25% range either side of the mean value (Table 1). Abundance of iron, zinc, and copper shows much greater scatter around mean values. High coefficients of correlation (> 0.995) were obtained for the various metals among duplicate samples of gypsum. Correlation decreases drastically in samples of size smaller than 1 g. This suggests point-form distribution of the metals in gypsum, perhaps as discrete mineral grains.

In contrast to the wide range of metal contents in gypsum layers, bottom muds show a significant correlation between metal content and organic-matter content. The organic matter was determined by loss on ignition at $1,050°C$ for 1 hr. X-ray analysis showed the ignited sample ash to consist of anhydrite. Therefore, the nonorganic fraction of the bottom sample was deemed to consist of gypsum. No clay minerals were identified by X-ray diffraction.

Metal content of the organic matter alone was estimated by extrapolating lines of best fit to 100% organic matter. Comparison of values shown in Table 1 with published values for bituminous shales in general (Wedepohl, 1971) shows content of copper, lead, and zinc to be at the upper end of or above the normal range of these elements, but still to be below maximum values recorded for the Kupferschiefer beds (Fig. 7). Unfortunately, very few data on trace-metal content of gypsum have been published.

The Kupferschiefer ore beds of copper, lead, and zinc have a spatial relation to evaporites, and the metals could have been concentrated in stratified heliothermal brines of the type found on El Roque. The stratification would prevent oxidation of organic material with which the metals are associated. Continued seepage or inflow of seawater into the system would have served to concentrate metals to levels in the Kupferschiefer. Lago Pueblo apparently is still too young to have reached such concentrations.

Conclusions

1. The lower-concentrated brine is enriched with respect to Fe, but is deficient in Pb, Zn, and Cu, if proportional concentration processes of these metals is assumed, or even processes calibrated on potassium enrichment in brine layers.

2. Density layering provides reducing conditions that preserve organic matter in relatively shallow water. Within the same order of magnitude, most of the metals missing from the concentrated solution could be accounted for by sorption onto the organic matter.

3. Base metals seem to be adsorbed preferentially or precipitated by organic sediments and removed from circulation. Iron and manganese—although also incorporated into organic sediments to a high degree—tend to be in brine in amounts greater than accounted for by normal concentration.

4. Iron in surface waters is exposed to amounts of oxygen adequate to form ferric hydroxides. As these sink into the oxygen-deficient hot zone, they are reduced to more soluble ferrous hydroxides. This may account for the excess iron in solution.

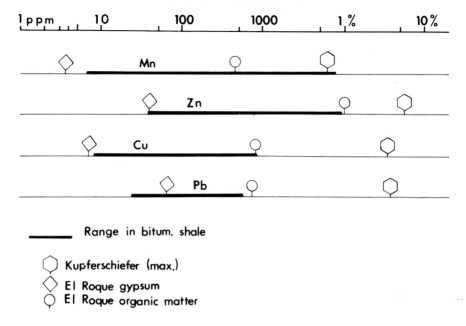

FIG. 7—Lago Pueblo samples compared to bituminous shales (after Wedepohl, 1971).

5. Iron, zinc, and copper appear to be distributed irregularly throughout the gypsum in discrete form, perhaps as specific minerals. Manganese and lead are dispersed more uniformly, suggesting atomic substitution.

6. Black organic muds are deposited immediately below the photozone. They contain values of metal greater than the range of bituminous shales, but less than maximum values of the Kupferschiefer.

7. Precipitation of gypsum seems to be restricted to the photozone, and gypsum replaces algal carbonates formed directly beneath the chemocline and thermocline of the lagoon.

8. Contrary to the consensus reached in a work group at the Saline Deposits Symposium in Houston (Ingerson, 1968, evidence for deep water basins, criterion 2), bituminous matter, sulfides, and reducing conditions are being maintained at depths of 1.5 to 5 m below the water surface; a model of deep-water deposition is not required.

9. A layered heliothermal brine model is suggested as the mechanism for base-metal concentrations in euxinic shale environments, such as that of the Kupferschiefer.

References Cited

Hudec, P. P., and P. Sonnenfeld, 1974, Hot brines on Los Roques: Science, v. 185, p. 440-442; v. 186, p. 1074-1075.

Ingerson, E., 1968, Deposition and geochemistry work session, *in* R. B. Mattox, et al, eds., Saline deposits: Geol. Soc. America Spec. Paper 88, p. 671-682.

Schubert, C., 1974, Striated ground on an arid tropical island; La Orchila, north-central Venezuelan offshore: Revue de Geomorph. Dynam., v. 23, p. 27-31.

Wedepohl, K. H., 1971, Geochemistry: New York, Holt, Renfrew & Winston, p. 131.

Reefs and Evaporites: A Summary[1]

L. L. SLOSS[2]

Introduction

At the October 1975 meeting of the Eastern Section of the Association it was not a burdensome task to summarize the results of a symposium on reefs and evaporites. The occasion for the summary coincided with the annual cocktail party and banquet and my contribution consisted of somewhat frivolous comments embellished by broad and sweeping gestures. On these pages, however, constrained by the requirements of formal publication, a summary paper that pretends to provide readers with organization and structure can not evade a responsibility for identifying opposing issues. Worse, the summarizer can not adopt a completely neutral stance and thus escape the displeasure of colleagues with whom he is in fundamental disagreement. Thus, I accept full blame if the paragraphs that follow seem less than evenhanded and appear to tilt away from a strict adherence to balance and equity.

The Bone of Contention

The preceding papers well illustrate the continuing controversy that has divided students of evaporites and carbonates for generations. Two schools of thought compete without significant communication from one to the other and without evidence of the emergence of a common intermediate ground. In over-simplified terms, the extrema of the two schools may be described:

1. Basinal evaporites are younger than enclosing basin-margin carbonates and represent the filling of a topographic depression developed by deep-water starvation of the basin interior, while banks and reefs flourished in shallow water at the basin periphery.

2. Basin-interior evaporites and basin-margin carbonates are nearly synchronous deposits representing coexisting but markedly different biological and geochemical environments.

We can refer to these extreme positions as the *batch-process* and *continuous-process* models. Each has certain well-defined tectonic implications. The batch process requires rapid subsidence of the basin interior to establish the environmental realms of circumferential carbonates and interior sediment starvation. Once the pattern is established it demands little reinforcement by continuing differential subsidence. Operation of the continuous process, on the other hand, would not be possible without fairly steady-state subsidence of the basin interior with respect to the margins, in order to maintain a near-constant differentiation of coexisting depositional environments.

The batch-process model presents a minimum of problems to geochemists and paleobiologists, and can be made compatible with the greater part of the observations taken from many local areas of evaporite-reef-bank occurrence. Further, acceptance of the batch model has been linked to and reinforced by the sabkha syndrome, which has approached pandemic proportions in recent years. Discovery of Holocene displacement sulfates on supratidal flats of the Persian Gulf emirates has led to attribution of all ancient nodular-mosaic, enterolithic, and chicken-wire anhydrites to prograding, supratidal environments at the margins of regressive seas. Sabkha-like anhydrites are common among basinal evaporites; if it follows that such anhydrites represent subaerial exposure of basin interiors, then sea levels must have dropped the tens to hundreds of meters by which pinnacle reefs and encircling banks stand above basin floors. "Evaporitic drawdown" is the process commonly invoked, and to strengthen their position, adherents apply observations of reef-derived rubble, vadose pisolites, and other evidences of depressed sea levels and supratidal exposure.

[1] Manuscript received July 21, 1977; accepted July 21, 1977.
[2] Northwestern University, Evanston, Illinois 60201.

Indeed, the batch process, synergistically combined with the sabkha syndrome, provides so neat and simple a model that as recently as a few years ago, it would have been a hands-down winner over the continuous process if the question were subject to settlement by a referendum among interested workers. The papers in this volume are remarkably evenly divided, perhaps signalling a trend toward more questioning attitudes. **Huh, Briggs,** and **Gill** come down hard on the side of the batch process, evaporitic drawdowns and supratidal exposure of the Michigan basin floor at the opening of Salina deposition. The Ann Arbor group waffles a bit in the paper by **Budros** and **Briggs,** allowing that not all displacement sulfates need be supratidal and permitting a degree of synchrony between Salinan carbonates and anhydrites, but holding firm to a demand for large-scale drawdown and essential separation in time of the bank-reef complex and basinal evaporites. **Friedman** and **Nurmi** agree that "displacive anhydrite could have formed subaqueously" but go along with the major tenets of the batch process for application to the Silurian of the Michigan basin.

The **Schreibers** and **Catalano** show that several biologic and depositional facies must have coexisted to develop the Miocene evaporites of the Salemi basin in Sicily. They remain faithful, however, to a Messinian "salinity crisis" and to the sequellae of such an event.

Other contributors to this volume, consciously or unconsciously, and in varying degree, support the continuous-process model in whole or in part. **Smosna, Patchen, Warschauer,** and **Perry** for example, interpret the evaporite facies of the Silurian Tonoloway Formation in West Virginia to represent basinal environments contemporaneous with carbonate deposition, the separation being effected by distance rather than by a discrete sill. **Davies** sees the Carboniferous Otto Fiord Formation of the Sverdrup basin as indicative of coeval basinal evaporites and marginal carbonates. Within the basin interior, carbonate-anhydrite alternations are viewed as successive high and lowstands of sea level, the latter representing modest evaporative drawdown that did not approach complete desiccation and development of sabkhas.

Matthews' exquisitely detailed study of closely-spaced cores in the Devonian Detroit River evaporites of Michigan demonstrates the continuity of individual units that pass laterally from carbonate to anhydrite to salt. Matthews notes the similarity of vertical and lateral facies change and suggests that, whatever the mechanism, single bodies of water covered the interior of the Michigan basin but varied in salinity and in depositional products from place to place. **Sonnenfeld, Hudec, Turek,** and **Boon** provide a hint of a mechanism by their report on a density-stratified lagoon in Venezuela, where chemically very different solutions and thermodynamic states coexist in one water body.

To me, the most compelling evidence in support of a continuous-process model is contributed in the paper by **Droste** and **Shaver.** Workers in Indiana and Illinois, whose areas of investigation include the divide between the Michigan and Illinois basins, have long been in conflict with a Middle Silurian carbonate - Late Silurian evaporite concept, but definitive data have been lacking. It now is demonstrable that growth of reefs and banks continued throughout much of Late Silurian time without major indications of catastrophic drawdowns or salinity crises. Thus, major volumes of carbonate were deposited at the margin of the Michigan basin in times that include episodes of evaporite deposition in the basin interior.

Conclusions

It seems clear that a retreat from broad application of the batch-process model to many major carbonate-evaporite complexes is in order. However, there are many stumbling blocks to be overcome before the continuous-process model or another alternative can be accepted blindly. One such problem is the subaqueous formation of gypsum and anhydrite by diagenetic displacement—a mechanism that is needed before the sabkha syndrome and extreme evaporative drawdown can be laid to rest. Even more pressing is the question of coexisting but seemingly incompatible chemical and biological environments. A solution was offered some years ago (Sloss, 1969; Stuart, 1973) but has not received wide popular support. Others (e.g., Mesolella et al, 1972) suggest a compromise achieved by relatively high-frequency alternation of evaporative and nonevaporative states, but almost as many problems are raised as are satisfied.

In short, the whole question of evaporite-carbonate relationships is ripe for a thorough-going reappraisal. Surely, a scientific community that survived decades of belief in fixed continents can develop new thoughts on the origin of table salt.

References Cited

Mesolella, K., et al, 1972, Cyclic deposition of Silurian carbonates and evaporites in the Michigan basin: AAPG Bull., v. 58, p. 3-33.

Sloss, L. L., 1969, Evaporite deposition from layered solutions: AAPG Bull., v. 53, p. 776-789.

Stuart, W. D., 1973, Wind-driven model for evaporite deposition in a layered sea: Geol. Soc. America Bull., v. 84, p. 2691-2704.

Explanation of Indexing

References are indexed according to the important, or "key" words. Authors and titles are also represented here; where more than one author has contributed to a paper, each person is cited, alphabetically, according to his last name.

In the column to the left of the keyword entry, the first letter represents the AAPG book series from which the reference originated (In this case, S stands for Studies series. Every five years, AAPG merges all its indexes together, and the letter S will differentiate this reference from those of Memoir Series, or from the AAPG Bulletin). The following number is the Series number (in this case, Studies Number 5), and the last entry is the page number in this volume.

A small dagger symbol (†) is used to highlight a manuscript title.

Note: This index is set up for a single-line entry. Where entries exceed one line of type (this is especially evident in title indexing), the line is terminated. The reader sometimes must be able to realize the keywords, although out of context.

Index